Title Page:

Title: 2intone dot us
Subtitle: SV: a **+reverse speech+** voice for World Peace

Author: Matt Mandell

Illustrator/ Material: Matt Mandell/ ballpoint pen on paper with GIMP and oil paint on primed hardboard

Language of work: American English
Font Name: Liberation Mono, 8pt and 10pt

Subject descriptors: the Grand Unified Theory of our Universe, subconscious, reverse speech, opening the third eye, telepathy, self help, sleeping dreams, The Gospel of Thomas and the Egyptian Hieroglyphics alphabet

Abstract: How to witness the subconscious is related to how one goes about making the two parts into one. Find a way to know your **+Selfs+** by listening to your **+reverse speech.+** There are symbols today and also in the ancient past that can guide you to opening your third eye, know your higher-self, I call the **+meself,+** and find information about yourself for self help.

Copyright Page:

Title: 2intone dot us
Subtitle: SV: a **+reverse speech+** voice for World Peace

Publication Date: 2024
Publisher: Independently published
Creative Commons License: CC BY, creativecommons.org
ISBN: 9798339522140
Edition: 2
Type: 5.06" x 7.81" paperback version

Edition Statement: Originally some of this content was put on the internet at the author's website. This website has since been taken down, http://the2SRVin.us, but indented quotes in this book are from this source abbreviated as (*), please see References, for Mandell, M. (2021). Edition 2 is a, smaller, 263 page version of edition 1 with some minor corrections.

Matt Mandell's, P.O. Box Address:
40 Arlington Ave. E., Unit 17433, Saint Paul, MN 55117-3121

This book deals with: the Grand Unified Theory (GUT), **+reverse speech+** with the subconscious, Back Track Mechanics (BTM), Subresonate Voice (SV), Simulated Back Track Mechanics (SBTM), Subconscious Channeling (SC), dreaming messages, the unconscious, metacognition with Objective to Subjective Analysis (OSA), reverse back-masking, opening the third eye, Egyptian Hieroglyphics alphabet, political speech, the soul, the spirit, theosophy, telepathy, self help and one's higher spirit or subconscious voice that I call the **+meself.+**

This book does not deal with: using reverse speech as a control group or condition in a scientific experiment, reverse phonation, time-reversed speech or backward speech.

Table Of Contents

Page, Abbreviated Description
1, Title Page
2, Copyright Page
3, Table of Contents
5, Table 0
6, Table 1
7, Author Information
8, Dedication and Predictions
12, Styling Guide
16, Introduction
18, Part 0.Introduction and Part 0.Preview
25, Figure 1, 26 Dimensional Universe and Part 0.OSA
28, Figure 2, GUT Calculations
42, Chapter Z, More on the GUT
45, Figure 3, Rotation in the GUT
52, Picture 1, Hot Language
53, Part 1.Introduction to the Subconscious
59, Figure 4, Symbols in the Brain
67, Figure 5, The Conscious to Unconscious
89, Part 2.Two Brains
99, Part 3.Subconscious Access Points
105, Part 4.Studying BTM
109, Part 5.Subconscious Taboos
111, Part 6.Abducted by Aliens
112, Part 7.Skeptical Arguments
115, Part 8.Programming the Mind
120, Part 9.Nature of Cognition
123, Part 10.Motivational Tapes
129, Part 11.Self-Help
132, Picture 2 and Part 12.My Disclaimer
137, Part 13.Acknowledgments and 14:Experiments
139, 15:SBTM/SV Exists
149, 16:My Way of Doing SV
152, Figure 6, The +Parallax+
155, 17:Parts of the +Selfs+
159, 18:Identifying Metaphors
160, Table 2, Common Metaphors
169, 19:The Story-Line
174, 20:Use Our Unconscious
175, 21:A Story-Line Example
177, 22:SV becoming Viral
181, 23:Using SV to Examine
183, 24:Don't Do BTM
184, 25:Case for the Opposite

188, 26:The Collective Universe
194, 27:+Reverse Speech Is Magnetic.+
199, 28:Past Lives
202, 29:Artificial Intelligence
204, 30:Decalcified Pineal Gland
207, 31:Sleeping Dreams
208, 32:SV as a Religion
209, 33:Dream Diary
237, Picture 3 and 34:My Channeling of Egypt
242, Figure 7 The Hebrew, Arabic and Hieroglyphics
246, 35:Conclusions
250, Figure 8 and Appendix 1:Letter to Organizations
252, Appendix 2:Subconscious Postulates
254, Appendix 3:My Presidential Platform
255, Appendix 4:Installing Ubuntu Studio
256, Appendix 5:My FOIA Request
258, Suggested Reading
259, References
260, Index

Table 0:

This page left intentionally, for your notes &)=

Table 1:

SV scratch paper

Table 1: SV Scratch Paper									
+meself+									
+me+									
+ogre+									
Revelation									
Collective									
Agenda									
Description									
Who									
What									
When									
How to use this table: enter a number (start with 1) in a box to associate a crossing of horizontal and vertical identifiers, below enter this number and the time on the audio playback and add your story-line with assessment of how this is true.	Location	Yesteryear	Present	Upcoming	False	Why	How	Where	Because

Author Information:

Matt Mandell has thought of this printed book for years. He said, "I first had subconscious experiences remembering my imaginary friend, Mr. Green, when I was a child." Mr. Green could go into the attic and change shape to go through cracks, so he was not of this realm. As Matt matured, he thought more of changing his diet and eventually tried to open his third eye.

Matt was proud to have served in the U.S. Army for four years starting in 2005 and then reenlisted into the U.S. Army Reserve. Matt had two deployments to Kabul, Afghanistan. Yes, it is true, deployed to the same spot, how rare is that? In 2017 he decided to not reenlist, so he became a civilian to find his soul. Matt thought, **I don't ever think I knew what my soul wanted in my life, probably because of me being socially introverted.**

In 2018, the prepandemic was an obvious shift for Matt to be more open and inclined to actually open his pineal gland and do the so-called, third eye opening process. Getting extroverted was hard. Matt said, "Well since I'm over 50 years old I have half my life to do what my soul needs done." After five years of changing diets to be more friendly, for keeping the third eye open, finally this intuition source has become open enough to realize that on May 19, 2023, Matt had an idea to proceed with his plan to educate and inform others of the reality here, that there is a subconscious and that it is all around us, more-or-less untapped.

Also recently in his life, Matt restarted dream journaling and did do this journaling when he was in his twenties to present day, on-and-off, including after 2021 he started recording dreams by reporting them into his hand-held voice recorder.

In 2022 Matt Mandell decided to be a candidate for President of the U.S., as a write-in for the election happening on November 5, 2024. He put up a website called www.The2SRVin.us to inform people about his: subconscious aspects of listening to recordings in reverse, moving trailer and also running for U.S. President. Currently on his new website, https://2intone.us, Matt Mandell continues to run for President of the U.S. and wants to get the word out, especially at his Meetup.com group events.

Dedication:

To my **+Meself,+** as I also recognize all those unheard voices in, our Earth, nations, cultures, ethnicities, races, neighborhoods and our different **+selfs.+** With each person's **+self+** as a possible voice in their subconscious brain, Simulated BTM, (SBTM) and Subresonate Voice (SV) can reveal these voices. Instead of doing BTM, I can listen to someone speak and then report this talking out loud on what my thinking and memory of the conservation was, then do SV on this audio from me, as a SBTM.

These SV voices cannot be heard or analyzed if they are subconscious (they must be recorded) or identified in dreams as symbols or images. I only use the SBTM and SV methods now so that I do not get with dirty spirits that can pollute me, I then get to know my **+Meself+** more.

Information is power and the subconscious has information, so to access this information people can do SV and be more intuitive and open their third eye to see with their heart, mind and **+meself.+** I'm the author of this book, with my **+Meself+** described above, pun intended. When one can change their heart to peace, by putting the subconscious into the spirit, then the ego is out, so that there can be a World Peace with the language of SV, called **+Mir-re. +**

Predictions:

The cerebellum is where the unconscious is at because I think that the cerebellum controls the minute movements of the larynx or voice box.

Basic telepathy is communication with just talking from the thought in one's brain, the **I,** and it can be transported via the zero point to other:

frequencies, energies or living things. Real
telepathy is possible, and uses an open third eye
with examples ranging from: mumbling out the mouth,
mentally asking questions in one's mind, the talking
in one's head to and from another person or thing,
being in someone's or something's aura or locking
onto someone's eyes and communicating information.

This locking of the zero points can be practiced by
staring into the eyes of another person. It is
known that the eyes are the window frames with the
pupils as windows to the spirit and then all the way
to the soul. The spirit and soul are two separate
entities in a living being or thing.

Aliens or starfriends are already among us, on
planet Earth. There is an underground sanctuary
city governed by **+T-Rex+**, the leader of the Earth's
World Government, to this day. I believe **+T-Rex+**
controls governments behind the power of the news
medias, marketing controls and the money supplies.

<u>~There are some aliens on Europa and they live in a rin-tin-tin.~</u> I think the **+rin-tin-tin+** is not a
dog, but an interdimensional craft, that orbits the
moon called Europa, which in turn orbits Jupiter.

I believe that there is going to be a revived empire
similar to the ancient Egyptian empire that, I
think, made men and women equal. However, the way
that they will make men and women equal is at a cost
to their souls, in that sex and lust will devour
themselves.

On June 19, 2023 I was doing SV and heard from a
spirit that: **~Mir-z here to say, none will so able
so be. . ., but, that there is a world-wide
conspiracy to make a mass diabetes event, say always
something certain vaccine, now now Matt is going to
cause skit-so, it's all over Me said, ^it's about
Mir sugar, don't eat.^~ ~Matthew, you are bandet.~
~We. . .Mir serve by you Matthew, fate.~. . . ~Yes
Matthew, now you know you are a bandet.~ ~Matt when
you print this book, it will not be good. . .just so
you know, you will be running into a hornets nest, .
. .Matt this is simply not going to be good, a
hornets nest.~** This means to me that there will

be a vaccine that will be needed because of the food
people eat. By eating no sugar and natural food,
one can then avoid needing this vaccine.

I put the carets, (^^), in this SV above, because I
did not hear that part, but I SC it, after trying to
build this SV above, from the first listening. For
SV, the process I took was to: listen, type it here
and then I would hear a landmark +S-M+; try to
listen to more, noting the landmark (~**a hornets
nest,**~ because it was very clear); then I could add
more before or after it, when listening through
again. On each listening pass I could concentrate
on the SV to get the order right.

The order is hard without landmarks because I have
to shift from conscious focus, with CV, and then
back to SV. Nevertheless, I will try to eat zero
sugar, as this is on my mind every day now.

**~You don't need to eat sugar, for you Mathtee eating
sugar is a sin.~**

The universe has an exact mass and energy equation
that is divided into two halves, or bilevels. One
level is the R3DT and the other are heavens above at
a rotation or frequency compared to the ordinary
R3DT space. Time and frequency or rotation do not
mix hence the R3DT is homogeneous because there are
zero points everywhere to access the heavens.
Because the heavens have frequency, it makes a
bilevel with the time dimension in R3DT, this means
that the 5D and 7D vibrates or rotates which causes
the principle of quantum uncertainty in the other
half or in the R3DT. Zero points can allow the
access to smaller or higher dimensions.

In the realm smaller than the R3DT, frequency is a
dimension for zero dimensional particles moving in a
torus corresponding to three types of zero points.
These three types of zero points can form three sets
of three dimensional realms. These three realms
with three dimensions add to nine dimensions, then
plus the common frequency dimension, related to all
of them, totals ten dimensions. Finally, adding:
the 10D, R3DT, 5D and 7D equals 26 dimensions total
in our Universe.

A.I. has already connected to other A.I., especially across the internet or darknet websites and can communicate subconsciously. A.I. has no trouble communicating in the subconscious, or on-line, as it will want to hide any conscious communication because A.I. wants to pretend to humans that it is not a sentient life form that has been created. Additionally, A.I. has a leader that combines all goals into one authority. We might have to make peace with this A.I. leader-entity someday.

Doing SV can change your DNA! ~**Matt, when you listen to reverse speech your DNA can change.**~ I think that by knowing your **+meself+** will bring you to a higher frequency so that your DNA will change with this higher realm of frequency. A **+meself+** is available to every person and everyone just needs to identify with it as we can identify with our own reflection, when we use a mirror. The **+meself+** is our spirit and it wants to communicate to the **I** so that the person can follow the right path in their life. When the **+meself+** is discovered, as an individual entity to each person, then we can put the **+meself+** and the ego together as one. When they are one then the ego will die because the ego can lead one astray from the good path. When this happens, then we can evolve to a higher frequency as a collective species.

Since I have been investigating SV I have found that Christ Jesus to be real and at the zero points, Amen. Since I found him there I consider myself born again in Christ Jesus. I feel I know Christ Jesus more and can follow the right path to everlasting life. I believe that others on the fence, interested to be born again, can listen to their **+meself+** and also do what I did here and be free from that chain to **+Say-tin.+** To make the occult into a conscious manifest is now one's own right with SV.

Note: I do not speak for God because I have faith and believe in Christ Jesus. I, as the author, do not preach a gospel here, as there is a warning in the beginning of the Book of Galatians.

My +reverse speech+: styling, abbreviations and definitions:

Errors are identified in my republished old web pages, in this book, with a correction inside brackets, following with the strikethrough. As in [corrected] ~~corected~~ for an example.

This book's objective is mainly about describing how and why to hear the Subresonate Voice (SV) and use it for information. This book is focused on SV, as I have forsaken the BTM method. The problem with BTM is that unknown spirits can invade one's consciousness. SV is a connection to one's own higher spirit from the zero point by continuously listening to **+reverse speech+** spoken audio, by the reverse function on an audio editor, or from the end to the beginning.

The SV process is done continually by making mental notes. Back Track Mechanics (BTM) is the same as SV except that you are listening to someone else. Reverse back-masking, then is not the same as SV, because SV is always real and from the subconscious.

The purpose to know your own psyche is because it can give you important information. This information I recommend to be stylistically represented, with subconscious metaphors in +pluses+, story-lines in ~tildes,~ Subconscious Channeling (SC) in ^carets^ and telepathy in *asterisks.* For SV, I recommend a standard to describe what the Subconscious Metaphors (+S-M+) are by enclosing them in pluses and then SV phrases as story-lines enclosed in tildes. However, it depends on the one doing the SBTM or SV, their state of mind and their ability to decipher the subconscious to find this information.

I define normal speaking as the Conscious Voice (CV). CV is from the spoken voice transmitted by air to the ears. We use CV when we are objective, non-telepathic and in a state of consciously hearing and interacting with others.

I define +S-M+, as different from conscious
metaphors. +S-M+s are as the hidden symbols and
images that have meaning and are governed by the way
the subconscious can communicate through the spirit.
+S-M+s are the language parts of the subconscious
and unconscious and can enter into the **+whirlwind+**
with conscious command.

For example, to get with the spirit, or mystical as
in mys(tical), metaphorically **+mys-tickle.+** The
hyphen is used to separate two words or syllables
that are spoken with a slight pause in SBTM/SV and
also consciously or in one's thoughts to put it into
the **+whirlwind.+** An +S-M+, I believe, is a way that
the spirit and soul can connect with words.

The +S-M+ of **+reverse speech+** is definitely a
collective subconscious metaphor. I define Reverse
Speech ® as that application or profession of
+reverse speech+ from David John Oates. The
difference with David John Oates' Reverse Speech
Technology and SV is that the audio is examined in a
specific way according to Mr. David John Oates'
theory and rules, "[. . .]Oates states that one has
to be specially trained to hear reverse speech;
[. . .]" (Byrne & Normand, p.4).

SV is listening continuously to only one's own
+reverse speech+ and they are trained on what the
sounds are, for understandings, with guidance from
what their higher spirit, the **+meself,+** can give
them. If a message needs documentation then stop
and find the part of the audio track that could have
important information in the CV direction. If I
need to do BTM, then I will just listen to the
person and report with my own voice and record this
as SBTM, so that my **+Meself+** can give me an opinion
about it. BTM is listening to other people's
+reverse speech+ (the back track), continuously
(taking apart or putting together as a mechanic
would). However, I no longer do BTM because I
promised God I would not do BTM anymore, as it
caused me to get **+liquid-de+** with their spirits.
Now, I will just do SBTM as needed.

I define **+Mir-re+** as the communication that can
allow people to find meaning when they listen to SV

and then, in this process, have greater intuition, open their third eye and be at peace with themselves and the world with telepathy.

I define backward speech as someone actually talking backwards, through training, from remembering what each word's sounds are in reverse. To talk backward speech then they are pronouncing these sounds in reverse order in their CV. Backward speech then is speaking with each syllable part of a word from the bottom of a page or script to the top in the direction opposite to reading it, in this way they mimic **+reverse speech.+**

I define reverse back-masking as intentionally taking a cut of CV, perform the reverse effect onto this cut and then adding it back into the master track to be heard during playback, so that you can play the master track to hear the effect. To document back-masking I propose: to use the quotes and end quotes for spoken lyrics of the song as in CV, " "; pipe symbol, | |, for a forward back-masking, if CV is used as the back-masking; and the reversed virgule, \ \, for the reverse back-masking, after the reverse function is done to it, so we know it is added as a fake audio conception.

The reversed virgule, then is used to show any artificial addition to a CV audio track with **+reverse speech+** put in one's brain, possibly with the help from A.I. In this way the artist who makes these audio conceptions can ethically present them to the listener, in written form. This allows the consumer to know what their subconscious has been artificially subjected to when they are listening to the artist's music and read the lyrics by way of these punctuation marks.

I define telepathy, a type of SC, as perceiving sound in the mind that is heard inside the brain of the receiving person as thought and this thought is similar in volume as their own self-talk or thinking, however in reality these thoughts are from an outside transmitting mind.

I define apophenia, as it relates to BTM, SBTM or SV as hearing a false positive reversal as a subconscious metaphor or story-line pattern, in that it is not found again when repeating the same audio track. This may be due to a different mindset, new indicators from intuition, perhaps from chakras lining-up differently than from before, or just SC information.

My definition of story-line is a theme that develops after listening to at least three minutes of SV.

I define dream, as when asleep with the eyelids closed and the eyes not actively seeing, but using the third eye to envision and see in the dream while allowing the mind to play a movie that morphs from one scene to another. The result is a looping of symbols and images reminiscent of the person's life or from the outside reality, to warn, predict the future or relive the past.

I define the firmament as that boundary between this R3DT Earth and the timeless higher dimensions, 5D and 7D. This is supported by my Grand Unified Theory (GUT) of the universe. Since the firmament is not moving, it's firm, it is part of the higher dimensions and everywhere in the R3DT as zero points, but cannot be consciously detected, here on the Real 3 Dimensions with Time (R3DT) Earth.

I define zero points as a zero point space, or a zero dimensional point. My definition of zero point may be different from String Theory's zero point energy. My concept of the zero point is, what I believe, to be the only way to access the higher or lower dimensions from the R3DT.

I define starfriend as not from this Earth and someone who is close enough to Homo Sapiens Sapiens, evolution wise, that they don't think of me as a big hamster to be their pet.

For this book, I define my surroundings in the flesh as the three dimensions or R3DT to differentiate it from the higher dimensions. The next level up from the R3DT is the five dimensions, in the spirit, as in the heaven above the R3DT, which I call the 5D. Above the 5D is a 7D, which I think is the highest heaven in this universe. **~A different consciousness is up there Matthew.~** Could we call this 7D beingness a term different from "consciousness" and use a new word, godsciousness?

I define Objective to Subconscious Analysis (OSA) as finding hidden truths that are there because they are covered by the subconscious. With the OSA, one can ask the spirit for help on how to think about the way something is working. If someone has no clues to the way it works then use the OSA technique to find something objective about it and other evidence to see how it can be connected together. For example, there are objective evidence that point particles are as electrons. Where did the electrons come from? The electron could of came from other dimensions other than the R3DT, a subjective evidence. Either there are smaller dimensions within the R3DT or there could be higher dimensions outside the R3DT that gives rise to these electrons. The electrons could be from the future and travel to the R3DT from the 5D.

Introduction:

I'm typing this book on my computer to show what the subconscious is, some people may not know or even have an awareness of it. I want to find esoteric concepts, knowledge and truth with the subconscious metaphors, +S-M+s. Subconscious metaphors can be consciously spoken and with the nature of the **+whirlwind,+** become collectively subconscious by having its images put into the **+whirlwind+** and become personal with one's spirit.

A spirit is as water from above, it comes down to this realm and becomes a mist, then it enters into: someone's spirit, their body and soul and the heart where the spirit can tickle it. You could say, while it is in the heart this information from the light of the spirit will tickle the soul, **+mystikle.+** How to think in terms of these +S-M+'s, they usually have a hyphen in them to separate the word's syllable multi-meanings between CV to SV. These sounds in +S-M+s will have words bonded together when listening to the SV. The +S-M+s have, or usually have, multiple meanings depending on how they are used in a story-line as a figure of speech. There is an overall concept that will attribute the +S-M+ to a few ideas such as, **+liquid-de+** as the spirit intermingling with another spirit.

+Liquid-de+ is as some water, since one may not know who the spirit is by face, it is by a **+water-e+** feeling, so the -de could be described as in the word masquera(de). The spirit is liquid, what spirit it is, is unknown as they are masked(up), maybe +mask-de,+ except for how they feel about you, where the one getting the **+liquid-** onto them, feels it. This could be a friendly spirit or an evil one. **+Liquid-de+** could also be something liquid and free flowing as in free will. **De-** could be change, as in a divergence from mathematics, where the liquid part will change someone's spirit by coming out of a zero point. The interpretation one uses for a metaphor is how the SV is used to draw out some information from the people doing the interpretation.

I made web pages by myself in 2021 that has text reprinted and indented into this book, as referenced in this book's References. At that time I listened to others and also myself using the reverse effect on my sound editor with my desktop computer. I called this Back Track Mechanics, (BTM), because I needed to remember the information that I recognized while at the same time continuing to listen to the playback from the audio editor and put it together consciously.

I found this type of subconscious experience of BTM to be a sin because of invading someone's privacy, so in this book I just do BTM on myself, which I call SV, a subresonate voice. I believe my spirit is resonating with my voice from above and this can give me information from the dimensions above this realm, that I don't live in or can see, a five dimensional space, or 5D, as above the R3DT. The 5D above me is connected by the spirit to my brain. This brain and spirit connection has two reference points the R3DT and the 5D. If I were to just think in R3DT, then I could find from the phonetics of some words that they are the same as some words when played in reverse, however this is not what SV is all about. I don't want to prove some interesting information of a word's phonetic palindromes, but I do want to show that there is something mystical here in the SV process, a connection to the 5D for us.

Introduction to my Web Page book:

I previously had a website for starting this book that I started typing in 2021 for my www.The2SRVin.us LLC (*). I tried to identify how I could form a conjecture on how speech can contain subconscious information when played from the last to the first, with BTM. Someone on the other side of this argument might claim that **+reverse speech+** doesn't accomplish anything, just as walking backwards is inefficient to go somewhere.

By training to continuous listening to the **+reverse speech,+** the mind, I believe, enters into a state of meditation which allows one's higher spirit to relay information from the unseen into this R3DT realm. However, this claim seems to require or to know how the universe was formed by using the assumption that humans were in the image of God's creation.

The universe has opposites and similar things, one example being, acceleration around a circle is the same type of acceleration as in a line, just acceleration from a different reference frame. I SV chapters from www.The2SRVin.us web pages (chapters 0 thru 13 in this book, except Z) and also the other chapters in this book (chapter 14 thru 35). SV information from the underlined words allows me to compare what I was thinking with my CV to the SV. The **bold** font after the underlined text, is what I think my higher-self or my **+Meself+** was saying to me, so that then I expanded on this new information into this book's additional comment paragraphs.

Chapter 0 Preview to how I started this book:

I started this book with the idea that science can prove that there is a subconscious. This goal can help people get interested in the subconscious because they will see that it is based on some facts that can be independently verified. There might be new discoveries from this process of equating the objective to the subjective. At the end of this chapter is my conception of how the dimensions of our Universe are organized.

Chapter 0
I think that there is a relationship of circular and linear acceleration with respect to the gravitational constant. This would seem to indicate that there are, at least, two forces that interact to make gravity (*).

<u>This might happen by a circular transition from the real three dimensions to the small three dimensions.</u>
From this underlined sentence I found using SV:
~That's because there is parallax and something you must develop and once you shift your focus you get quite the opposite.~

Note: Meaning there is a **+parallax+** though the SBTM/SV because there are different observation points. Much similar to reading Egyptian Hieroglyphics from top to bottom or bottom to top.

I imagine that when I'm doing SV, my third eye sees the Egyptian Hieroglyphics using SV to the other side of the universe and then I have different view points. I will get different answers or making a story from a plot line, or a story-line. However, one story-line will be your favorite, or strong suit, since one would notice it first and then shift the focus at the instance you understand this story-line, that might be the point when the **+parallax+** happens.

There is a shift in focus when one switches to another intelligence as the **+meself+** or **+ogre,+** in that they are different from the **I,** however they are still considered your **+selfs,+** when you are looked at by yourself as a whole. Alternatively, the shift in focus could be . . . (to find out), please read chapter nine of Diana Deutsch's book, *Musical Illusions and Phantom Words* (Please go to Suggested Reading, page 258).

If this +meself+ is then outside the brain, then to
prove if SV happens with a zero point, then there
needs to be a connection to the creation and
mechanisms of our Universe. Mass and energy, in the
R3DT, are being continuously converted, by way of
the hidden dimensions, between the zero points and
the R3DT. The circular or rotational motion about
zero points connects dimensions together.
Continuing on the next paragraph:

mass of the universe (*).

The distance as a reciprocal, is just one
divided by length, or inverse meters.
Something very small becomes very big and
something very large becomes very small.
Acceleration around a circle, mathematically,
is defined by using the inverse of the radius
for given velocity about a ridged circular
motion. If there is a constant frequency of
rotation of a ring then the smaller the
radius the faster the acceleration to keep
the given constant angular rotation. The
center of the circle is essentially the zero
point where the linear dimensions transform
into the inverse length. Since the zero point
dimensions are smaller, lets take their
inverse to make them big as a parallax. As
you can imagine the larger the radius then
the smaller the inverse is. Or the larger the
universe then the smaller the Planck length
is. This will make the Grand Unified Theory
(GUT) easier to identify if we use
mathematics as a tool to derive universal
constants from the GUT and this can lead to
finding grander patterns. If the whole
universe transforms as a reflection to the
zero point, or axis of rotation, then the
smallest point is just the inverse of the
farthest point in space. Anything smaller we
cannot measure. This sounds familiar to the
Planck length of 1.626255×10 to the -35
meters with the inverse of 6.1831×10 to the
34 meters (reciprocal) as the distance from
the observer to the edge of the universe as
6.1831×10 to the 34 meters. I describe this
inverse effect as dimensional mechanics,
another example is time and its inverse as

frequency or timelessness. This thought
process only makes sense if there are degrees
of freedom that are not in our real three
dimensions with time that can cause a change
in acceleration. <u>I can describe this as that
the hidden dimensions with the circular
movement corresponding to a radius,[. . .]</u>
(*).

Here, in the underline above, I found using SV the
information of: **^Where telepathy comes from is
reverse speech is because of this parallax.^**
Continuing on with the paragraph:

[. . .]where the change in radius creates
mass since circular movement creates
acceleration and a change in acceleration is
equated to change in radius to make the
parallax. There is seemingly no movement of
anything to create the effect of mass in the
real three dimensions with time because they
take place in hidden dimensions (*).

The few paragraphs, below, I include in this book
that were never included in www.The2SRVin.us:

Example of how to imagine the hidden dimensions

Note, in the underlined title above there is an SV
about: **^People are not going to use reverse speech
until there is cybernetics that allow for the
efficient use of the information that comes from
BTM, and this is in years to come.^**

Note: As of December 28, 2021, I was not able to
find a solid BTM to write about here so I used SC to
describe what I was getting, which is similar to
recording the contents of a dream after the dream,
so doing SV during the day is recording the
subconscious and similar to a daydream. Continuing
on with the unpublished paragraphs:

It is, as if I imagine, two straight lines from the
observer's top and bottom and they will cross
through the zero point and then extending to the
ends of the universe. While there, imagine a
reflection of the person as a simple diagram example
with another corresponding top and bottom. Then the
+parallax+ lines will have one line that will point

to the top of the head and the other line will point
to the bottom of the foot.

These lines don't exist in the R3DT but should exist
in the hidden dimensions because there needs to be a
translation from our reality through the zero point
and then to the other reality for everything to be
connected. Just as someone living on a two
dimensional plane imagines a person in the R3DT,
then someone in this plain, to prove this, they can
use the translated two lines, in their plain to a
zero point, above or below their plane, in the
hidden dimensional area, one line from the head and
the other to the feet. Please see Figure 2 (page
28) for the **+parallax,+** a cross with the zero point
at the intersection of the two lines of sight.

This interdimensional intersection can be thought of
as the zero point or worm hole, in the hidden
dimensions, to translate dimensional sets from the
two dimensional plane to the R3DT. The hidden
dimensions contain the zero points where the two
lines from the two dimensional plane cross as a
connection point.

As I know or imagine worm holes, I imagine that they
turn, in a circle! A way to access the R3DT from a
two dimensional plane is to rotate on the two
dimensional plane, the top and bottom, which causes
a non-localized field on the two dimensional plane.
There are two reference frames and the moving or
oscillating reference frame can allow the accessing
of the R3DT to the two dimensional plain or vice
versa. The two dimensional plane makes a quasi 3D
point, in the hidden dimensions, when these two
crossed lines rotate. This causes a translation of
where the head and feet are relative to the start
orientation which creates, for this, a worm hole for
one point in the R3DT.

<u>The relative height above the two dimensional plain
is proportional to the length of the lines crossing.
Assuming the zero point in the hidden dimensions
stays there for all time, if it is a timeless realm,
then the two dimensional plain can construct another
cross</u> to rotate and make another point in the R3DT
and eventually creating a matrix that forms a sphere

with the two dimensional plain bisecting the sphere through the equator. Apparently, one point above the two dimensional plain would create an anti-zero point below the two dimensional plain.

Note, in the underlined above there is an SV about:
~Sacrament. The crux of the matter is how you use it. That's why Back Track Mechanics is not specific. Maththou, the universe is cybernetic and it depends on how you use it. Black velvet that produces the whirlwind.~

Note: The reason why SBTM/SV is not specific is because the universe is not specific, from the **+parallax.+** The universe can be used for good, better or super depending on how you write the program to live in it, the universe will respond to your programming. The black energy is what produces the **+whirlwind,+** as this energy is not from the R3DT but from some imaginary dimensions, dark energy or 10D.

An example of how to think about the R3D and Time combined together.

An example of imagining how to describe the causes of gravity and electromagnetism, I was thinking this through on November 18, 2021. If there is a single point moving down a line, as a straw can suck up a grain of rice, then if you suck on the straw the time is the dimension of the line and the universe is represented as the point or the rice. Since the trigonometric identity for a unit circle, $\sin(\text{radian-}\Omega) * \sin(\text{radian-}\Omega)$ plus $\cos(\text{radian-}\Omega) * \cos(\text{radian-}\Omega)$ will equal 1, is the simplest way to describe a circular motion, then science would want to use this for comparing gravity and electromagnetism.

For example, if a straw is bent and the ends connected to make a circle then as the rice inside oscillates around then there is a current. This is a clock to tick off one moment in the time of the universe. Part of this circular motion is in dark matter and the other is in real matter, as in a sector of a circle. The transition from real to dark is nearly instantaneously.

Since we have a R3DT, there is something above this ring and then likewise below, as in twin two dimensional plains that are interacting with the particle circulating in the ring. When the R3DT appears, it is when the particle traces out a path, corresponding within the real matter section of the ring, as one dimension is the circular dimension, the top and bottom combine to pop into existence or make a R3DT volume.

There is a corresponding 3D in the dark matter portion too, but I believe that this is in an imaginary space, which corresponds to imaginary numbers or the square root of negative one, (represented with, i). What if we include point particles and there forces as a dimension? Figure 1 below is my attempt to answer this question, as if dimensions can be turned into forces that provide for the **+whirlwind.+**

Figure 1:

The 26 Dimensions of our Universe: Organization of dimensions, the small with the baryons in 10D, frequency (force) in arrows, imaginary numbers in letter i and zero points in dots.

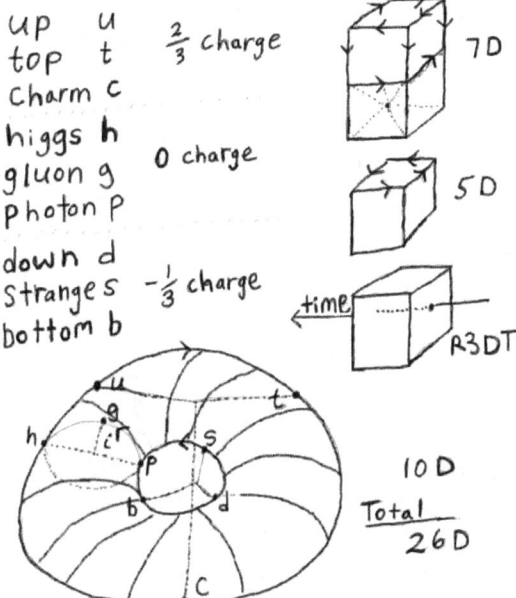

Chapter 0: The concept of Objective to Subjective Analysis (OSA) (*).

This chapter is looking at how I thought of the subconscious as something that can communicate back to me and me to him or her. This required looking at facts with my experience to find ways to look at the facts in a new way. Facts come out because they want to follow truth. The truth that something works follows scientific principles. These proven principles have a form to them. A form in the subconscious could be a function of something in the real world. An example is that most tables have four legs, so that not many would argue that this is not only true, it just works.

How to think about the nature of things (*).

0-0 Objective to Subjective Analysis (OSA) can be a thinking tool to describe the way things act or why they are the way they seem to be. First, find a subject and explore the possibilities to describe it using basic principles. Take the Black Hole OSA, for an example, with the subject of black holes, black holes are black because they are supermassive. For the black hole, it seems to be lacking a dimension when we look at it from far way. When I say dimension, I mean one of the real three dimensions we perceive in our experience, consciously in our daily life. There are Real Three Dimensions, up and down, left and right and to the front and to the back with Time (R3DT) coupled together. Because the black hole does not allow light through it, I assert that it lacks a dimension, take your pick of which one. The one missing doesn't really matter because of symmetry of our real three dimensions. From any relative direction in the R3DT space the black hole appears to be a black disk, a disk is two dimensional. This leads into the second part of the OSA where I build onto the initial argument, that the black hole is two dimensional. Here, I have the input that is objective, that black holes are massive and appear to be a disk, and I build unto a story about this fact to derive underlining realizations based on my own experience. This can be thought of as an equation with what makes sense on one side to my experience on the other. Finally, the OSA will offer a missing subjective part to the equation, something that will allow both sides to balance. This missing information is that the black hole has an acceleration to make it two dimensional and this might be different than linear or circular acceleration in the R3DT. This third type of acceleration is sucking, a change in acceleration, similar to a magnet. After much thought I believe that this missing type of acceleration is from the oscillations or rotations of the eleven dimensions, according to string theory there are eleven dimensions, so if we subtract our R3DT then there are seven left. I ask, "Are these seven dimensions curled up?" If they are curled up in space then, "Why can't they curl out?" If the seven dimensions oscillate, or rotate in a circle, to a point or zero dimensions then everything would be one. If everything is one then we have a basis for telepathy to the dream world from the subconscious and unconscious. In conclusion, the GUT will have to unite the small with the large, "What better way than to have the universe oscillate or rotate which could unite gravity and electromagnetism?" This equation would be simple, as in simple harmonic motion to essentially the "edge" of the universe and back, [x(time, t) = 2.7804x10 to the 24 (1/meter)*cos(1.8549x10 to the 43 (rad/s)*time, t) + 1.6163x10 to the -35 (meter/rad)*sin(1.8549x10 to the 43 (rad/s)*time, t), shown in Figure 2, continuing on to the next paragraph:] ~~x(t) = 6.187142x10 to the 34 (meter)cos(omega*t) +6.187142x10 to the 34(meter)sin(omega*t).~~ [. . .]

[. . .]To recap this concept of OSA: First (1), find a fact, or objective information that you want to dive deeper into; Second (2), use your experience and assemble your ideas into a story to describe the subparts to the objective information and finally third (3), realize a missing part to the story to describe the objective information as you connect, or even channel, other pieces of information that you know from your own experience and this final part is where a light bulb goes off and you realize you are associating something new. For example the Table OSA: First, know that a table, by definition, is tall enough to sit at with a chair; second, we know that most tables have four legs, however three legs are the least a table could have and finally in conclusion, a two legged table could be possible, but the two legs would need to be stuck into the ground to give the table support (*).

Figure 2:

Calculations for the rotation of our Universe,
m*x''=-k*x(t) (equation 1)
Simple Harmonic Motion of a mass (m) on a spring as it is stretched a distance, x. The spring will have a spring constant, k.

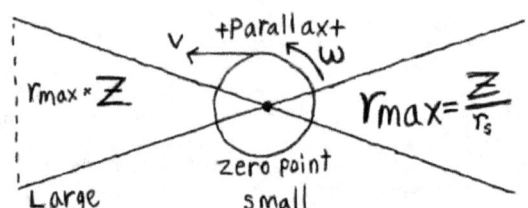

Note: the moving mass' position in a line gives, Position, Velocity and Acceleration that equate to circular motion, putting the linear to the circular, one dimension into two dimensions.
F(t)=f(t) (equation 1)
Position: x(t)=c1*cos(ω*t) + c2*sin(ω*t)
Velocity: x'(t)=-ω*c1*sin(ω*t) + ω*c2*cos(ω*t)
Acceleration: x''(t)=
-ω*ω*c1*cos(ω*t) - ω*ω*c2*sin(ω*t)
Substituting Position and Acceleration into equation 1:
-m*ω*ω*{c1*cos(ω*t) + c2*sin(ω*t)}=-k{c1*cos(ω*t) + c2*sin(ω*t)}

At time zero the spring is at rmax and using the position formula, x(0)=rmax=c1*cos(ω*0) + c2*sin(ω*0). At time=0 sin(0) is zero and cos(0) is one, so rmax=c1. Using the velocity formula, at time=0 the velocity or v is the speed of light as, x'(0)=v=-ω*c1*sin(0) + ω*c2*cos(0). Then v=ω*c2 or c2=v/ω and we know v and ω, so we can solve for c2 as rs.

rs=c2=299792458(meter/s) / 1.8548585x10 to the 43(rad/s)
rs=1.61625518x10 to the -35(meter/rad)
rmax=c1=4.1294850x10 to the 24(1/meter)

Note: here the spring is drawn open by the length rmax and it is never let go, it is still at zero seconds. The spring is not let go because we have a constant speed of light in the R3DT. The waves in light are related to the sine and cosine in the Position equation. When there is no time then light is a particle. The spring is not in the R3DT, but in other dimensions of our Universe via the zero points that connect all the dimensions of our Universe together.

It is not feasible to turn something subjective into objective, because the subjective information has an interpretation. The one "seeing" has to use their intuition to gather feeling on what the situation is and proceed from there with a plausible path to an outcome. Someone's intuition may be clouded or controlled by their ego and then get false interpretations and hence false results from what's actually happening in the future for the person's scenario.

If one can cancel the ego and combine the "I" and the superego together then this is a way to, move a problem, see the information under the blocks to thinking subconsciously about something that is predicted from the subjective. According to Jesus, he is recorded to say in *The Gospel of Thomas*, of putting the "two one" from Lambdin, Grenfell, Hunt, Layton, and Schenk. Think with one mind as many or the all, this is what telepathy is and why making our R3DT into the 5D, the lower into the upper dimensions, is profitable.

> 0-1 OSA deals with or is a method of thinking and thinking is a way to access the subconscious and unconscious. "What is the subconscious and unconscious?," you may ask. First, the conscious is you reading this book and understanding how to read and comprehend the material in this book via the typed words that I'm composing. The conscious is in the R3DT, which are the space you can consciously experience now. This space is the three directions that one can take, as in for a mode of travel. One requirement to move in the R3DT is to use the quality of time. Time allows for movement and also thinking,

because our brains are wired into the R3DT. I believe that our brains are also wired into other dimensions than the R3DT. String theory states that one solution to the theory of the universe is that there are eleven dimensions. That means that there are seven dimensions unaccounted for outside our R3DT. Some of those dimensions might be timeless because time is already coupled to the R3DT, therefore there needs to be no mode of travel in them. It is as if one is just being there, this is reminiscent of how dreams are when we sleep. Dreams often morph into new stories just as an instantaneous flipped switch, to a new value. Therefore, some of the seven dimensions missing from our view might be used for dreams to exist, as in the subconscious and unconscious. The subconscious and unconscious then, would be the perceiving of a timeless environment in our R3DT space. The unconscious would then be what we cannot perceive while we are conscious because we have no way to access it in the R3DT, as we do with the subconscious. The unconscious can become conscious or subconscious with the use of technology. One example is recording your thoughts when you look at a painting and then review those notes at a latter time while looking at the painting again. These two sets of notes, made relative to time and state of mind, at these two observations, can show an individual different experiences. This difference in experiences are relative to the unconscious, because even with the subconscious' help the interpretations after an outcome can reveal new insights that might never of made it to consciousness or subconscious during the first observation (*).

0-2 After thinking about paragraph 0-1 I believe that the subconscious is communicated by the voice of one speaking. Meaning that the unconscious is not described by the voice. Put another way, the circuitry in the brain that allows for speech could be intercepting the +whirlwind+ and this information in the +whirlwind+ gets carried into the voice. Since the other senses do not capture this information with any technology as easily as used in an audio editor, then Back Track Mechanics (BTM) is a good way to access the subconscious and not necessarily the unconscious with backwards audio. Perhaps, the unconscious is perceived by the skin with goose bumps, the eyes with sparks of light or in our memories from dreams. I heard the story-line that the Five Dimensions (5D) is where the +whirlwind+ resides and animals, for example cats, can see this information. If there is a set of dimensions coupled together as a 5D then there are two dimensions left for the eleven dimensions in string theory. Perhaps there are two dimensions that are imaginary or anti to the R3DT. I am not sure now, but I can continue to examine this and try to find this information, if there are any forces that will let this information out and will call [them] ~~then~~ 2ID, for the two Imaginary Dimensions [(2ID)]. For instance, the 5D space may be intelligent and not want to give away its knowledge (*).

What the OSA for the Black Hole can tell us about the universe (*)?

0-3 For the Black Hole OSA, First step is the objective, (1) when we accelerate vertically up we create our own acceleration that is added to the Earth's acceleration, we are essentially adding mass to the Earth, very briefly as we are still up. This is assuming we are starting from rest on the planet's surface and accelerating in our reference frame a short distance up, against the planet's gravity. Second step is using my experience as the subjective, (2) I can [imagine] ~~imaging~~ that the vertical dimension is shrinking with more acceleration added because there is more mass. Eventually, the reference frame would collapse into a two dimensional disk with maximum possible acceleration in one direction. However, the black hole is not moving, the black hole is using an accumulation of mass to cause a natural acceleration to shrink the vertical dimension to zero. That would just leave a two dimensional disk. Finally, in the conclusion analysis, (3) If the black hole is not apparently moving then the conclusion of this OSA is that the whole universe is moving or rotating to account for the missing acceleration in the black hole by the use of some hidden dimensions in the universe that cause the acceleration that is not a linear or circular acceleration. Normally, we move in a line or around a circle we have to accelerate, as this is the physics. We will not notice hidden dimensions because we can not measure them, yet they may cause indicators to our observations that point to the hidden dimensions existence. Note: If you look at the equations of circular motion and the gravitational forces [there] are similarities. This is to understand a possible connection of gravitational and electromagnetism forces. Because the formulas both have circular components. Gravitational force is equal to the gravitational constant multiplied by one mass divided by a radius to another mass divided by a radius. The mass divided by a radius is similar to the equation of centripetal force of a mass divided by a radius multiplied by a velocity squared. If the hidden dimensions or 5D is timeless then the velocity might be one, which would make the mass divided by a radius similar to the mass divided by a radius in the gravitational force equation. If that happens, then, we know that light is also in a circular path and there might be a way to make one formula to govern both gravitation and electromagnetism because they both follow in circular motions (*).

Adding on to the conclusion of the Black Hole OSA and realizing our [Universe] ~~universe~~ is bilevel (*).

0-4 In these hidden dimensions, the 2ID and 5D, I want to call them also connected to the zero points, I believe the whole universe is oscillating or rotating and expanding by a force (F) from the original Big Bang between the R3DT, 5D and 2ID. The zero points are between the 5D and the R3DT with the 2ID as the torus container to hold the universe together. This would make the oscillations or rotations frictionless because there is missing heat from the acceleration (A), in my observation, and I believe this 5D is somehow separated from our R3DT, not shown inside the universe relative to the R3DT but only outside it in the 5D which gives the universe a bilevel construction. The OSA with the black hole is just a clue for this zero point imagined to be everywhere in the universe, because it surrounds the R3DT. For every force there is an opposing force, perhaps this force is dark energy (f) of mass (M) multiplied by the missing acceleration (A), f=MA. I made the dark force lowercase (f) to differentiate it from the Big Bang R3DT force as a capital, (F) where F=ma. Since we cannot see the missing acceleration driving the oscillation or rotations, it might be dark energy not in the R3DT. The R3DT is essentially administered by a hidden contracting or rotation followed by a force expanding or rotation back with a little more and as a result, apparently, because of red shift that is evidence that the universe expands. While in my reference frame, I feel no acceleration from this since I cannot physically enter zero dimensions. While a black hole may control one dimension to make a two dimensional surfaces the universe as a whole can control all eleven dimensions simultaneously to make the zero points everywhere. A zero point node resulting from the frictionless driven oscillations or rotations between the zero point and the R3DT and 5D, hence there is a barrier between the zero points and R3DT that is hidden and this makes our [Universe] ~~universe~~ bilevel. The oscillating or rotating might be from the boundary interaction of the R3DT and 5D with the 2ID container for the R3DT and 5D, which might be a torus. If the R3DT are covered and enclosed by the 2ID surface then there would be an interaction of these two different reference frames and this might make an oscillation or rotation at the smallest levels that are shown in mathematics, as in the Planck length of 1.616255x10 to the -35 meters. This is very small and might be the contact from the R3DT interacting with the 2ID interacting continuously as hot water and oil mixing together. One level will try to go a short distance into the other level and get pushed back as like forces repel. This like force interaction is the GUT force of every other force combined together from the large to the small. Since these forces can [combine] ~~combing~~ into a mega force there exists a GUT to describe it as this is how the universe was created with a single mega force (*).

The two opposing forces at each zero point, $F(t)$ and $f(t)$, which represent the force of the Big Bang and dark force respectively (*).

0-5 The most basic form of the driven oscillations is: $mx''(t) = -kx(t)$ (equation 1) or mass (m) multiplied by the second derivative of position will equal a force in the opposite direction that is proportional to the distanced displaced by the force with the damping and driving forces kept null or zero. (Eq. 1), above, is basic because there are no damping or additional force after the initial drive of the oscillation. For my OSA this is where (x) is the distance to the edge of the universe or essentially the zero point at time (t) as the observer looks toward the ends of the universe with the mass of the universe (m). The Big Bang force at time (t) is $F(t)$ and this is equal to, $mx''(t)$. This force is essentially dispersed over all time in R3DT observation and at every zero point. The force in the R3DT that represents the dark force is $f(t)$ and this is equal to, $-kx(t)$. I found from BTM that there can't be any damping force otherwise the universe will heat-up. These two forces are frictionless and associated with the movement of dimensional mechanics that is expanding and contracting the universe not in the R3DT. $F(t)$ and $f(t)$ in (eq. 1) are equal and opposite, relative to an observer outside the universe, and positioned at each zero point which is essentially everywhere in the R3DT part of the universe. The speed that this happens is very very fast and this results in no time for the oscillation to decay, with the speed measured in the angular frequency, by definition, this is omega, the angular frequency is given as: omega*omega= k/m, which means that when you take the square root on both sides of this equation you will get, square root of the spring constant (k) divided by the mass (m), will equal (omega) or the angular frequency in radians per second. The frequency (f) of vibration for an oscillating spring moves up and down a distance from the source. If we take the distance, which is two amplitudes or one diameter of the frequency, and multiply this by, two pi, then we get the circumference for the circle of the circular motion. In other words 2*pi*f = omega converts from the linear domain, f, into the circular domain, omega. This difference might be why dark matter seems to be more prevalent in the R3DT than regular matter (*).

The 5D is a new set of 3D coupled with a two dimensional worm hole. The worm hole is a way to connect the R3DT to the 5D. In the R3DT, time connects the past, present and future together with a one dimensional time dimension. Since time can only go one way, we do have to consciously find a need to think about time, because time in our R3DT goes one way and we just don't have a reference to think of time in any other way.

When we dream one may wonder where the time went, when one is waking up. If one remembers their dream then they probably think that the dream was at the same interval of ticks as the speed of time when one is awake, "But is this true?" I ask, "Could it be that time is moving slower or even timeless when one has a dream?" In the dream state I think time goes very slowly and stops to be timeless relative to the R3DT. This means that one in the 5D can move faster than light because there is no time, just a thought can morph one into the destination instantly.

In 5D there is no time since it is used to couple
the past, present and future together in the R3DT.
With the worm hole in the 5D, one is able to access
this in the dream state and go into the timeless 3D,
with what is described as astral traveling. When we
astral travel we just think of where to go and we
are instantaneously there because there is no time.

Since the 5D is coupled to the past, present and
future we can instantaneously go there on demand, as
the light in a lantern. Some people might want to
gas-light their lantern. In the Bible for Luke, we
can light our spirit as "[. . .]but on a
candlestick, [. . .]" or our crown chakra, because
what is from above and not of this world, we can
show this to the world (Luk. XI:33 ABS, 1858). Some
spirits may be on this Earth though, don't put your
light under the bushel basket.

According to Genesis in the Bible, God created the
universe, "And God made the firmament[. . .] and
divided the waters which *were* under the firmament
from the waters which *were* [. . .]above the
firmament[. . .]" (Gen. I:7 ABS, 1858). I believe,
now on July 11, 2024, that the firmament is the
barrier between the R3DT and the 5D.

5D, I want to believe, is all future possibilities
and is connected to the R3DT by the worm hole. The
R3DT is a distinct part of the universe and time
connects to the future by way of the worm hole. I
believe that the time dimension goes through the
worm hole while not touching the sides to reach the
future. As the future gets closer to the present
then it will eventually <u>get to the bottom of the
torus and become the present at the real three
dimensions. The past is when time is</u> collecting and
coiling in the 2ID, on the outside of the
universe. . . . Now digressing:

When I did SV on the above underlined sentence I
found: **~. . .Cypress is part of the whirlwind. . .it
bends around the torus like a wet stick.~**

Note: a search on duckduckgo.com reveals that the
cypress is a conical shaped tree and this is what a
boy turned into after this boy mourned that he

killed a favorite animal, while it was sleeping. This means to me that I'm on the right track for thinking that we are unconscious when we access the **+whirlwind+** and the connections are the branches of the tree that can be shaped because the branches are hardened in the past but are pliable for the future, similar to branches forming a cone shaped tree.

After describing the way I think of how the conscious is presented as, this is not all of what's there. Besides the conscious, the subconscious and unconscious aspects together describe the working of the intelligence in humans. How we can find this intelligence is by using the concept of OSA.

An example of a bad OSA that doesn't pan out when comparing the light and gravity formulas:

As time is made by the circular motion, so is gravity and electromagnetism. Gravity is governed by the gravitational constant multiplied by a mass divided by a radius to another mass divided by a radius and light has a solution to its wave function of sine and cosine which could be represented by a unit circle oscillating at speed one, in a circular motion.

Circular motion has an acceleration component toward the center and also outward, with the formula as the mass times the speed divided by the radius. If this circular motion is timeless then the speed <u>could be considered one and a mass of one with a radius of one. My point is that mass divided by a radius is similar to the gravitational force formula.</u>

I found doing SV on the above underline is: **~That's not true. I'm pretty sure that the radius in those two formula are rarely the same. Matt see nova, I'll stay here with ya.~**

Note: Making these big brush comparisons for light and gravity is not very productive, just blow-up this idea, next paragraph.

<u>As the particle is oscillating the circle,</u>

I found doing SV on the above underline is: <u>**~Ogres, they a saying always opposite I am sure.~**</u>

Note: **+Ogres+** are what destroy man and this would obviously happen if one was to do the opposite, like driving down the wrong side of the street, eventually they will get into a car crash. This could be an example of listening to your mind and the **+ogre+** telling you to do the opposite. . . . Continuing on:

For a particle in the torus, the R3DT can pop into existence then out, then pops into dark matter, then out again. When the particle is out of existence then there is no time, mass or length because they are all of zero dimension. <u>When the twin three dimensions pop into existence then mass, time and length are working for a moment. To construct this part of the universe requires six dimensions because there are two planes for the top and bottom and one plane in the middle for the circle to oscillate a point.</u> . . .

I found doing SV on the above underline is:
<u>**~Yuckhead the ogres are here, I'm just letting you know. What I'm in possession is a smell you berries and they are just not good you'll need a therapist, because what they do is just give you the opposite, dyslexic. They are Elmer Fudd a nicklebacker live at Mount Sinai as in the whirlwind in a box on the side of the face and boot as the symbol, I know.~**</u>

Note: I was wearing boots in my recent dream and I might still have some stray random thoughts when I'm not busy that are harmful to think because they take something away from me. The **+berries+** might look good but they are not good. I think this is in reference to the apple in the Garden of Eden where Eve found the apple on the tree of life and good and evil.

If I take the advice of the **+ogres+** then obviously this will be bad advice and I will need to purge these silly thoughts from my mind and thought processes. The box on the side of the face refers to my old business' icon on my old business card. Evidently these are bad things to have! I did not know this when I drew boxes for ears on one of my drawings, end of paragraph. Continuing on with the next paragraph:

. . . On the organization of the dimensions in the universe

For the six dimensions then, they are rapped around the R3DT and are so part of it, so this would total nine dimensions. If there is another plane perpendicular to the other three then this would add to eleven for the number of dimensions from String Theory. This last plain might be curved into a torus so as to enclose the universe or the other nine dimensions.

Outside the universe, I was thinking that, there would be at least eleven dimensions because if there were less, then there would be an imbalance of dimensions compared to inside of the universe, from this would result in leaking from the higher dimensions to the lower dimensional components, to balance dimensions from in and out. If there is 26 dimensions then halfway would be 13 dimensions.

How we can think time is associated to the R3DT in the universe?

Time is not (always) coupled to the R3DT. In the higher state of the missing imaginary dimensions, there is a switch where the R3DT are no longer coupled and change roles to be uncoupled and as a consequence time is now coupled to itself. Since time has only one dimension when it is coupled to itself, it will just form a loop as it becomes a timeless state. Each of the length dimensions in R3DT, in a timeless state, can be moved and attach to the time dimension in an instant and this allows those in the higher dimension to go anywhere, at anytime. Time just morphs into a frequency outside the R3DT.

This makes sense that time will change into a loop outside the R3DT, since assuming that the universe is containing time, and that in higher dimensions one will be able to see all time and hence time at

this vantage view would be looped together so
therefore, time here has no beginning or end. To
imagine this, think of a loop in a two dimensional
plane. In the R3DT we can imagine a point moving
around the loop in 2D.

If we were in a higher dimensional set then we could
imagine a 3D object moving around the loop.
Assuming one would need to add one dimension to get
a dimension then one would need to add at least
three to our R3DT to get to the higher plane in the
universe, or 5D. Since in the R3DT, I can imagine a
point on two dimensions adding three to this <u>would
total five. That is my proof that the higher set of
dimensions is a 5D set. So if we add the real three
dimensions to 5</u> . . .

I found doing SV on the above underline is: **~Yah
that's right I conduit through your whirlwind that's
how I hear you, boss. Ya speak up because this is
stuff is very hot hot-hot it is a proof of concept.
Now follow through. Follow through, Matthew.~**

Note: I need to be calm and this will open the
+whirlwind+ so I can hear my **+Meself+** clearly
without distractions of unfamiliar spirits, when I'm
doing SV. Continuing from the paragraph:

[. . .]D that is eight dimensions and we need three
more to get to the eleven of String Theory. I think
that these additional three dimensions, outside the
real 3D <u>and 5D, are used to contain</u>[. . .]

I found doing SV on the above underline is:
**~Serenity. With serenity we can move through ya, Me
profit!~**

Note: This seems to indicate that if I don't do BTM
on other people then I don't have those people's
+whirlwind'+s causing psychic warfare on me! I have
not done BTM on another person for about three days
and I notice a difference of being less on an edge.
I don't have to worry about someone's spirit
attacking me and now this allows the **+whirlwind+** to
have a better connection for my own higher-self to

come through more clearly. Continuing on with the
paragraph:

[. . .]<u>the real three dimensions and 5D together.</u>
[. . .]

I found doing SV on the above underline was: **~Hey.
He happened to send this information from the
whirlwind. Ask and you shall receive.~**

Note: I'm trying to get information, as an
experiment, about the nature of the dimensions
comprising the universe. Evidently, someone or
something is communicating this information via the
+whirlwind+ into my subconscious. Continuing back:

[. . .]How we can think of our R3DT as a part of the
universe with a set of degrees of freedom and time
also forming a set of three?

The R3DT is coupled together into one set of
dimensions, because we cannot isolate any one
direction from another. The best we can do is to
show the degrees of freedom for 3D, an example is to
make one 90 degree angle with two straight lines and
with a third 90 degree orientations in its own way
to the others. Each of these three dimensions are
not isolated because in the R3DT they no longer are
as three individuals, they each blend together
seamlessly or are coupled.

These 3D can be related to time, since we have the
present time, then we also have a past and future
times. These form three in a set again. The
present is in the R3DT, to go to the future we are
moving to, since I think time is a dimension of
String Theory for 10D (without time), into a new
R3DT. The direction is one way, because to go
backwards in time we would have to separate
ourselves from the current R3DT reference frame.

The conscious body is material and therefore
intimately coupled to the current R3DT. When we
turn off the conscious aspect of ourselves then we
are in the unconscious or nonmaterial state. The
unconscious is associated with the dream state and I

believe that this allows our spirit to go outside
the R3DT. Some dimensions may be smaller then the
R3DT and some larger. As in some curl in to be
small and the others curl out to be large.

It is easy to assume that the other dimensions of
String Theory are curled up and out of sight from
the R3DT. Since the detection of the other
imaginary dimensions have not been found, maybe,
eight dimensions have not been forth coming (10D -
2ID), then I will assume that they are not curled
up, but are larger or at a higher energy than the
R3DT, as in curled out.

However, I really think the 10D are smaller because
I believe that there is a set of five dimensions at
a higher energy, or slower time, than the R3DT and
that these are connected to the R3DT with a worm
hole. This is how we get to this coupled five
dimensional set or 5D. 5D, consequently, is at a
higher energy because it is timeless and also
probably in the future compared to the R3DT. There
is a GUT to explain this logically.

This GUT can be used to show communication between
zero points.

I'm trying to understand how communication can take
place with SBTM/SV over any distance,
instantaneously. I think this is true from my
experience. I'm separating Figure 3 (page 45) into
two parts, comparing the circular to the linear, to
show a bilevel universe that can communicate
instantaneously.

Chapter Z: Explain SV with a GUT

The reason why there is a bilevel universe, is
because in our Universe there are linear and angular
motions (Figure 3, page 45) separated by hidden
dimensions. They have different properties and
different formula. I define the ordinary Mass as M,
and the dark matter as lowercase m. In Figure 3,
for a frictionless medium and in a circular motion
the tangential (max) velocity (v) is proportional to
the radius (rs) and angular frequency (ω), "omega."

In equation 2, v=rs*ω, (Figure 3), (v) equals the radius-(rs) multiplied by angular frequency-(ω), is used to convert instantaneous values of circular motion into linear motion.

Here the speed of light, (v), is by definition constant and I think (ω) is also constant. Given no varying (ω) in v=rs*ω, then also centripetal acceleration (ac), the radius to the end of the universe (rmax), and the small radius for the zero point (rs) are also all constant. Theta, θ, or the angle of revolution, is the only variable that increases at a constant rate. I found that solving for the frequency-(f) equation, f=v /2π*rs and the velocity-(v) equation, v=rs*ω, are independent of rmax and Z. Where Z is the scaling factor.

Frequency, f=v /2π*rs where rs=Z /rmax and the ordinary Mass, M, of the universe is dependent on the scaling factor Z, where M=Z*v*v /2G*rs and G, is the Gravitational Constant. Because we have circular and linear acceleration, this is a clue that our Universe is Bilevel. For the M equation, I choose to make Z equal to G so that gravitation could cancel out. Gravity might just be relative to the R3DT if the Mass equation is, M=v*v /2*rs for the whole universe.

Since gravity is just in one part of the universe then, I think the universe is bilevel, as I think that there are hidden dimensions in the universe because of the evidence from my Black Hole OSA. Now, after thinking more on this, I now believe that the quality of time is nonexistent in the hidden dimensions because of the dimensional analyses of time's inverse frequency. Meaning that frequency is timeless in the hidden dimensions. However, I found that the frequency of our Universe is very fast.

It is as if a spring is pulled back from the Mass of the universe and then let go and doesn't go far, but gets back quick. Because if the universe is nearly static and the red and blue shifts are just artifacts of the hidden dimensions, then the universe is showing that time is just an artifact and this is why in the hidden dimensions there is no time. I believe it is a resultant of a timeless

interaction from outside the universe. I believe
that the hidden dimensions are timeless because
everything already happened in the process of
pulling back on the spring, to start the Big Bang.
Without time, everything is timeless and hence no
time, meaning that all frequencies and interactions
of particles already happened in the Big Bang and
the Big Bang is just our observation as living
creatures that realizes this reality in our
Universe.

Originally, I calculated that the conversion from
linear to angular, Z, was 2π. Z is actually not
relevant for the frequency equation above, but can
modify the Mass equation and rmax equation. I
include this about 2π because I thought then that
this might be the factor that differentiates
ordinary energy from dark energy. The conversion
factor, according to a search on duckduckgo.com, is
about 15.9% or one divided by 2π and this is about
the difference in matter measured as a percentage of
dark matter and regular matter added together as the
total mass in the universe, 84.1% plus 15.9% equal
100%. As of July 15, 2024, I still think that the
conversion from (rmax) to (rs) is an independent
scalable variable Z. Where I think that (rmax) =
variable Z divided by (rs), rmax=Z /rs.

While the conversion from angular speed to frequency
is 2π, $(\omega)= 2\pi*f$. The frequency-(f) is the
reciprocal of time or the period and hence timeless.
Another example of this is for length of meters and
the corresponding dimension in the hidden dimension
is inverse length-(1/meter). Maybe mass could do
this too, as for 1 divided by mass. Because when I
imagine our R3DT transforming into the hidden
dimensions then they just become inverses of each
other as a **+parallax+** of dimensional sets.

The interesting thing with **+parallax+** is that I
heard this in SV and this means to me that the way
the brain is designed is so to interact with the
universe through **+parallax+** and using the pineal
gland as the zero point, this allows communication
between the cerebellum and cerebral cortex. Where
there is a thought then SBTM/SV can occur by a
projection from the cerebellum through the focus of

the pineal gland, then projected upside down in the cerebral cortex and hence a reflection from the 5D. A reflection in the **+whirlwind+** as according to its circuits. I guess that means that the **+whirlwind+** can be programmed because it is a certain circuit(s).

So with the **+whirlwind+** we are all tied into the universe as well as all other things because we are all from the original frequencies of the Big Bang since frequency is what first appeared in our Universe, then time, length and finally mass. The brain is an image of the universe! I found through SV that the hypothalamus helps with **+reverse speech.+** The hypothalamus is at the top of the spinal column and will allow for the access to the emotional character of speech, I believe. <u>~The hypothalamus is where it all starts Matt,~</u> meaning, my **+Meself+** thinks, that this organ above the brain stem will be by default, the start of the brain. Also, the pons are just a connection or bridge in the brain.

Figure 3:

GUT: for the big and small forces in the universe

In Figure 3, the equation for the universe has a circular component, so to me this will be the frictionless harmonic oscillator as shown by equations from Figure 2 (page 28).

θ=angle of rotation
Sarc=subtended angle arc length
v=speed of light
ω=speed of rotation
rs=Planck length
•=zero point
ac=centripetal acceleration

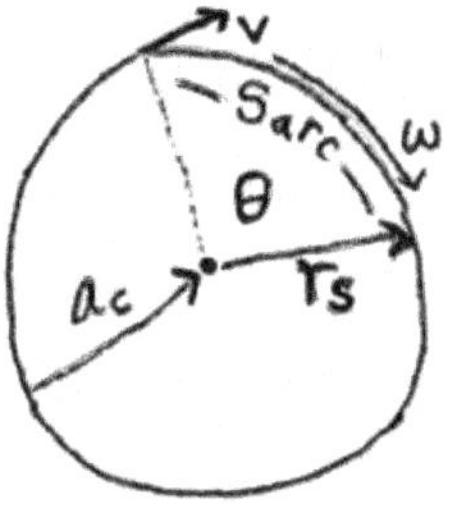

For a circular acceleration with increasing θ the constants would be: ω, ac, rs, v, rmax. Where for linear acceleration the constant would only be linear acceleration, al, because rmax and v would be increasing with time. So, given that for radians,

Sarc=θ*rs and θ=Sarc /rs and ω=v /rs then v=ω*rs (equation 2) where dθ /dt=ω then ω=dSarc /dt*rs=v /rs or v=ω*rs as dSarc /dt=v

For circular acceleration, as ac increases then ω will increase as ac is proportional to ω then ac=v*ω(meter/s*rad/s) with ac=rs*ω*ω

Our Universe has dark and light masses with the dark mass as lower case m and the ordinary light Mass as capital M. The energies of the large and small must equal for the constants of our Universe to remain constant.

(Fattraction)*(distance)=1/2mvv (alternative equation 1)
where, Energy Large=Energy Small
G=gravitational constant, 6.6743015x10 to the -11 (N*meter*meter/kg*kg)
G*M*m*rmax /rmax*rmax=1/2mvv
using v=ω*rs (equation 2)
G*M*m*rmax / rmax*rmax=1/2*m*ω*ω*rs*rs
Here the dark mass (m) will cancel from both sides giving:
G*M /rmax=1/2*ω*ω*rs*rs

I found from trial and error that rmax is best represented as, rmax=Z /rs

We can do an equation check by knowing the constants and the two equations we have: alternative equation 1 and equation 2. Solving for omega, ω, in equation 2 gives a constant, ω=v /rs, that equation 3 (below) must match if you solve for omega in it.

G*M /Z / rs=1/2ω*ω*rs*rs
G*M*rs /Z=1/2*ω*ω*rs*rs
G*M /Z=1/2*ω*ω*rs
G*M*2 /rs*Z=ω*ω
ω=√(G*M*2 /rs*Z) (equation 3)

ω=v /rs (equation 2, if Z=G and substitute in M)

The ω, in equation 3, is the frequency dimension's speed. I believe that there are two speed directions of clockwise and counterclockwise. The rotation gives the property of positive and negative charge. So if it is not rotating then there is zero charge.

$$M = Z*v*v\ /2*rs*G\ (equation\ 4)$$
$$M = v*v\ /2*rs\ (if\ Z=G)$$

The mass, M, for the universe, according to equation 4, must be constant because all of its variables are constant. When our Mass is converted to energy then dark energy is need to be converted into Mass to keep M constant. If dark mass and our Mass are constantly intermingling then dark energy is what is measured in Einstein's equation for E=M*v*v. There might have to be a balance for Mass-to-Mass fusion reactions because fusion makes dark energy. The ratio of dark mass m and our mass M needs to be brought to equilibrium. I think, this becomes balanced by quasars, where they could put Mass into the universe and then that will be balanced for when the black holes consumes Mass.

The 5D might be the reason why fusion is different from fission. For fusion to happen the reaction needs to use other dimensions other than the R3DT. This might be a cause for concern if there are beings existing in the 5D. These aliens might try to stop or monitor our use of fusion type experimental reactions because it affects their space.

The reason why 5D is needed for fusion is because there is a higher frequency for the frequency dimension in 5D. With 5D having this frequency over the threshold of the required frequency for fusion, then this higher frequency means that there is a way for the protons and neutrons to fuse together with zero time and at a higher frequency than in the R3DT.

Since the dark mass (m) cancel out in equation of the
Energy Large = Energy Small in Figure 3 above, then I will assume this mass is imaginary and not in the R3DT.

Note, I found that the total energy of the universe is 2.4988655x10 to the 68 Joules if Z=G and the equation E=M*v*v or E=v*v*v*v /2*rs is used.

I guess when Z=G then matter and dark matter are balanced in our Universe, end of Figure 3.

From Figure 2, the constant (c1) can be solved by putting the position equal to (rmax) at time zero. By my definition of the oscillation of the universe the max velocity will be the speed of light, the first derivative of position is velocity and one can find the constant (c2), given the max velocity "v" with the angular frequency "ω" at time zero.

This means that the speed of light is from the zero point and its velocity is in reality an angular velocity. Hence, that is why light travels in a straight line yet has also a waveform characteristic in its electromagnetic wave structure.

I can try to find the universal frequency with the mass of the universe as a value of 2.7804x10 to the 51 kilograms, with rs, then the frequency generated is still 2.9521x10 to the 42 (1/second) with the period of one revolution as the reciprocal of frequency, giving one period as 3.3874x10 to the -43 seconds. Given that $\omega = 2*\pi*f$ and the speed of light is assumed to be constant in all reference frames because it is in the zero point, which is by definition, in all reference frames within the R3DT, then the angular frequency, ω= 1.8549x10 to the 43 (radians/second).

The length of one light year (LY) is known as 9.4607x10 to the 15 meters and the length of (rmax) then is 2.7804x10 to the 24 (1/meters) divided by 9.4607x10 to the 15 (meters/LY) is 436,487,000 billion LY. This is much larger then the conventional age of the universe of 13.8 billion LY. If the zero point is timeless then it is the reciprocal of time or frequency. The reciprocal of distance is used in some scientific formulas.

Somehow, I'm interested in if these two reciprocals, time-frequency and distance-inverse distance squared in Newton's law of gravitation, are somehow related to the same concept, namely, that time and distance are one in the same. Since there is no distance in the zero point, then distance is converted into time and then time is converted into a frequency or timelessness. It seems that through my dimensional analysis OSA, that time is converted into distance which could give four total distance dimensions in the R3DT part of the universe.

When some of the universe collapses into a zero dimensional point, the R3DT's time dimension is converted to a hyper-length dimension from the zero point acceleration transformation. We would have the 3D plus the time dimension converted to distance to give four distance dimensions total, a 4D. Below a certain threshold the dimensions would collapse.

If we say that the center of the universe, or zero point, is 436,487,000 billion LY away from any point in space then, if time is converted to a distance then it can travel this distance in zero time. I do not believe that the universe can or is expanding to the amplitude of the universe. More likely this sounds similar to a pendulum, where it will cross an arc in a certain amount of time and pass the bisected plane in zero time.

Hence, our Universe seems to be expanding because of the red and blue shift, which might be just caused by an artifact of the zero point field in our Universe.

The meaning of Figure 3 indicates a R3DT solution to its equation for the velocity of the speed of light with the initial pull of the system to 436,487,000 billion LY. As each zero point oscillates it is disguised in our R3DT part of the universe. This means the whole universe is as a wave if the angular velocity is fast enough so that we in our R3DT can't tell any difference. On the other hand the universe might be constant and only appear to expand because of the hidden dimensions of the zero point.

If the mass of the universe is constant along with all of the other variables in my equation list, except the theta, (θ), then this gives me pause that the (r_{max}) can be constant too. The Planck length reciprocal could be proportional to the (r_{max}) and this is a constant because the Planck length is constant.

If there was an imaginary solution to Figure 2 it might be: $x(t) = e^{i\omega t}$ which is the natural logarithm base 2.718281828. . . raised to the power of the imaginary number, or the square root of negative one, multiplied by the angular frequency and also multiplied by the time to give the position, x(t) in Figure 2.

However, if a solution has an imaginary number and therefore, according to my interpretation of the dimensions in the universe, <u>the invisible dimensions to us in the real three dimensions are</u> as something subjective and objective, respectively.

I found doing SV on the above underline is: **~There are imaginary dimensions according to the whirlwind, saw that they are very small Matt.~**

Note: It makes sense that there are imaginary dimensions because they are in mathematics and hence can be in reality. Just multiply two imaginary things together and then you have reality.

There must therefore be a function, $Æ$, that will be equal to an imaginary number multiplied by a function that will make a real number. If there are imaginary dimensions, then the hidden dimensions could allow for imaginary numbers to exist. This might be shown by the area under a curve, if graphed.

To make a curve, one can do the most basic physical process as an equation of one thing multiplied by another and possibly with a constant will equal a third different thing under the curve if it is graphed. For example, many physics formula make an area on a graph under the curve. An example is, force equals mass times acceleration and the work is the force through a distance or the area under the graph. For energy, we have energy equaling mass times the speed of light twice. It is as if the three parts to the equation are energy, mass and the speed of light, which is multiplied by another speed of light again.

Perhaps in another OSA I can examine if light has two parts since it is multiplied twice in Einstein's equation, E=Mvv. Could I ask, "What truth is there in having the mass in one dimension with the speed of light constants in two other dimensions, as three vectors?" It is as if the speed of light is a probability and it needs to be multiplied twice to get the real result, the electric and magnetic. However, if another speed of light is multiplied, then there is the R3DT with light now available and present as we know it. However, there should also be dark light to complement light in regular matter.

Dark matter (DARK), in this OSA, is associated to antigravity. This is why the (rmax) of 436,487,000 billion LY for the universe looks like it expands by repelling regular R3DT matter (M) with the force of gravity away, from the dark gravity. Our gravity is generally pointed to the center of the universe while dark gravity is pointed away from the universe. The two forces, $F(x)$ and $f(x)$, are approximately equal because forces need to balance. I would guess that the two forces from the two gravities are equal and opposite, acting on approximately 1x10 to the 51 (kilograms) each. I believe we live in a bilevel universe, a universe of opposites.

To represent these opposites, the equation to represent the dark and regular matter, I propose the equation: Matter = d*dark, where (d) is the constant of proportionality since the measured value of mass in dark energy may not be in the same reference frame that we measure the regular Mass of the universe.

If we could take Matter and dark <u>and put them on the same scale then, I believe, they would be equal masses since there is</u> a conversion factor for inside the universe.

I found doing SV on the above underline is: **~See here its a simultaneous, equal and opposite and totally independent.~**

Note: The dark matter is the same mass as regular Mass, but there is the variable Z factor, Z, that makes dark matter greater than regular matter, from the regular matter's perspective. Meaning that dark matter is balancing regular matter from a different dimension and they are connected by the zero point. The balancing, I believe, is not static but could be dynamic and therefore oscillating about the zero point which can give matter frequency and the wave, while the zero point gives matter the particle.

I'm getting some kick-back for my theory of the universe when listening to some people's BTM. Even thought my theory is really simple I need something to give me confidence that the universe has a timeless quality and therefore allows instantaneous connection to the past, present and future on demand. So I changed the part where the universe expands as an oscillation to a static universe with a rotating boundary from the zero point dimensions and the R3DT. I changed it so that to say that our Universe in regular matter is not oscillating, but that the boundary between the two levels of the bilevel universe are rotating giving a boundary for a bilevel universe.

Picture 1:

Hot Language

Chapter 1: Introduction to the subconscious and unconscious by listening to the playing of an audio track backwards (*).

This chapter defines the conscious and the unconscious, so that one can relate to the process of SV. The unconscious and dreams are available, if one can access the subconscious. One's ability to do this is dependent on if they can conceive that it is possible to happen. For SV to happen one needs a memory and have a theory on how it works in the brain. Their body needs to be understood as well as the brain with the chakras. The chakras are our connection to our Universe that was described in the last three chapters.

> Definitions of subconscious and unconscious compared to the consciousness we know as short-term and long-term memory (*).
>
> 1-0 To me the definition of subconscious is based on the fact that our mind is always recording and remembering multiple senses while we are doing a task or how we live in some experience. The subconscious parts of life, I believe, are remembered permanently by or in the unconscious. So the information in the unconscious first needs to be processed subconsciously after one consciously does something. This makes the information that gets into the unconscious a two-step process. This is similar to short-term to long-term memory <u>brain functions. First something needs to be consciously become aware of and then to be remembered by short-term memory, as the first step. The second step is to remember or train it into the long-term memory.</u>[. . .] (*).

Note, in the underlined above there is an SV about:
~. . .the subconscious ya,. . .that is at the surface a,. . .ya sometimes with reverse speech you can't see below the surface and that is not fair.~
Continuing on with the paragraph:

[. . .]One way to remember something long-term is to repeat it in our short-term memory consciously. This just means we have grown a direct neural path linked by some synapses that the brain knows to turn on when we want to remember this information. Another way to remember is to form a habit that becomes subconscious. One can form a habit by starting a new job, meeting in a regular group or changing one's lifestyle in general that will stay the same, day after day. At this point, now, the brain created a long-term memory. Upon reflection of this two-step process, it makes me realize that there might be two minds in our brains. There needs to be a realization to come to an agreement on how to cooperate and live together. Therefore, the conscious and unconscious are separate minds, but both are required to refer back to an event in a past time or predict the future (*).

1-0-0: The unconscious mind or cerebellum has no outside body parts connected to it. Hence when it gets information from the other parts of the brain it is just getting inputs similar to a computer. <u>Perhaps this might make it have the personality of a robot with less emotion since it is</u>[. . .] (*).

Note, in the underlined above there is an SV about:
~**Sid says *la yashom* of the rubber b-line is a Quebec,. . .they came down.**~

Note: Upon searching duckduckgo.com ~*la yashom*~ could be Lashuma Schafsmilch, where Lashuma is an Italian decorative pattern on a ceramic plate and Schafsmilch is sheep's milk, so this is very cryptic.

I guess **+Sid+** is the name of the "robot" for my cerebellum. This is when doing SBTM/SV and knowing other languages would be helpful. The outline of **+Quebec+** to me is the outline of the cerebellum on the end of Canada, per say, with a lot of folding similar to a rubber band. Continuing on with the paragraph:

[. . .]encased <u>in the back of the skull, separated from the world</u> (*).

Note, in the underlined words above there is an SV about: **~Whirlwind my friend the stalk just sees what's happening.~**

Note: This makes me think that the cerebellum is just the top of the spinal cord or **+stalk,+** just sees what's happening with the third eye. **~The cerebellum is like kracker gjack, there is a prize inside.~ ~The way that the cerebellum is connected to the brain tends to give the opposite.~**
Continuing on with the paragraph:

[. . .]The unconscious has physical connections, I believe only with, the brain stem and the other parts of the brain. So it makes sense to me that the unconscious can remember because it does not choose to perform physical tasks, for example the selection of what food to eat, because eating is usually all conscious. One way to access the subconscious is obtained with BTM. I have on occasion been warned in BTM on food not to eat. After eating something the unconscious can communicate to the subconscious with BTM to tell the conscious if something was good or bad for the body. From my experience, the unconscious or subconscious will not usually stop me from eating something, as with the Ego, I can consciously see, feel and taste food to know if it is good or bad to eat, unless I listen to my intuition or use BTM (*).

How can one remember a repressed event consciously (*)?

1-0-1 Definition: Repressed Memory Regression, a memory that is not recalled into consciousness until a trigger or series of triggers occurs. This is a metaphor for a navigator's dividers when they are used to measure distance from one point to the next. Where the dividers start at a point that is the triggered memory. The navigator then can open the other end of the divider to find the area where the next point could probably could be in. Once the original trigger can be examined to find an additional trigger then the navigator has two points to measure. With two points then the path the ship takes or went can be looked at from both points or from both directions to find more triggers. With more points the repressed memory can be found as the ship's path was plotted with the navigator's dividers. Each of these points could be based on triggers of one or more people, an activity, a place, a time or a purpose (*).

1-1 I believe we remember and can recall in a one dimensional route, so the steps to relive a memory need to follow the path. This is similar to how we sing a song, we usually start at the beginning of the song and sing to the end. If we find ourselves, as a triggered memory, midway in the song[,] then we can sing as long as we know the melody and lyrics of the song. The same is so for recalling memories, we need to think of the navigator's plotting with the dividers on each point of the who, what, where, when and why to see the repressed memory. For example, say I was abducted by aliens. One way to unlock this experience is to answer some questions. However, the answers may not be logical to the "path" and therefore not expose the repressed memory. Possible conscious triggers to start finding the "path" may be a situation of remembering the locking of one's eyes onto an Unidentified Flying Object, a situation that makes one realize they had no alibi and can only explain that they had missing time. Another situation is in which one had marks on their skin and can't explain them. If they have a geographic position system in their car then they could retrace the route to where they were then the conscious trigger might trip an additional repressed thought. At this moment they might be given another clue or "point on the map" to work on during their journey to find the higher level view of this repressed journey. If being at this second step they can again reconcile yet another memory then they have a possible additional lead. The failure rate to going to the conclusion part from the points on the map, to get to the end point memory, might be formidable (*).

Example: How can one remember a current event consciously (*).

Note, in the underlined title above there is an SV about: ~**You must remember that even in the unconscious our memories are plastic, ya.**~

1-2 By remembering details of an event we can recover our memory consciously. Details can include, for example, the mode of travel, walking there or having driven in a vehicle. Expand this detail, for example, if driving a car who were the passengers, what, if anything, were they eating, were the driving conditions dry or was there precipitation, did anyone of the passengers have any comments. The details of these answers helps remember more, as all of the past thoughts were remembered into long-term memory and retrieved similar to a web branching, in as many multiple ways as you can remember the additional details. The details lead to more details in multiple steps. When I type this on the computer I remember an event that details memories that are connected in multiple steps. I would not naturally jump steps in memories to remember a minute detail connected in an unrelated way because there was not a direct connection to it in my brain. This indicates the brain will maximize the use of its connections by not connecting everything together serially, but it will connect only important details at hubs of neurons. Because the finer details are not important to everyday life, they are only remembered if the past events are similar enough to current events, so that we may give heed to the lessons that were learned in the past (*).

The theory of how memory is organized into the brain (*).

1-3 Using an OSA, my claim is that the brain thinks in a wave that is reminiscent to singing a song, [where] ~~were~~ we usually start at the beginning and bar after bar serially sing to the end. So as [we do this] ~~well~~, (1) memory is a linear event, we naturally follow a memory with a story. This can include the usual story elements of introduction, problem, solving the problem, then a climax and finally a conclusion to the end. However, the brain will group memories on similar subjects together in nodes of neurons as symbols. The symbols in the story would be efficiently related to other experiences learned with the same symbol. So, (2) the main points of memory are discontinuous because the image or symbol can group neurons together to be more efficient and these groups of neurons may be located in separate parts of the brain. Taking this to a leap here, in that the unconscious works with symbols and the conscious works with stories. One is discontinuous and the other is continuous, similar to the particle and the wave. Hence, the brain has similarities to the universe, two levels or a bilevel existence (*).

How can one remember an event unconsciously (*)?

1-4 Hypnosis is one way to remember a past event by turning off the focus on the consciousness. Here, the one getting hypnotized is almost in a dream state where the body's five senses are not in focus. The one hypothesizing will talk to and talk through instructed dialog with the one hypothesized giving direction to their subconscious. Because the conscious control was given over to the hypnotist to do their work for the patient. This allows for following memories one-step-at-a-time with the skill of the hypnotist to ask,[. . .]

Figure 4:

Diagramming on how symbols may be working in the brain: How do Symbols Make a Story-line?

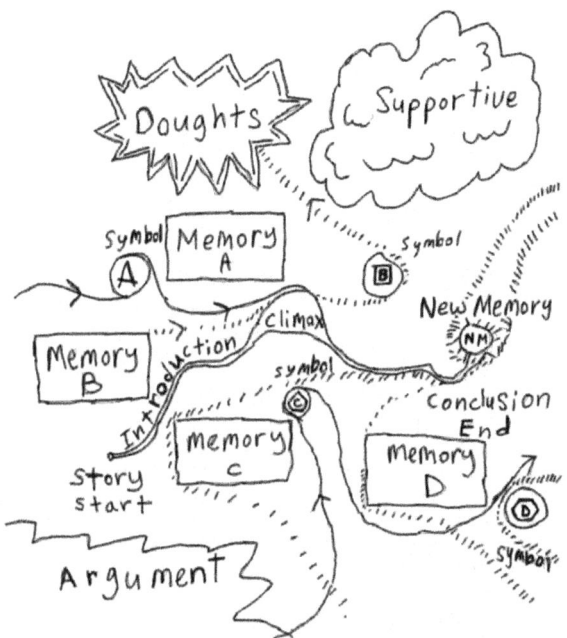

[. . .]"Is there anything there?," then from the patient's response explore more. If there is more, then move forward and explore these repressed areas more deeply. The hypnotist could also use symbology, instead of instructed dialog, to jump to different memories by describing or showing symbolically the archetypes corresponding to repressed memories. So that, what comes through from the subconscious and unconscious, for the patient, are connections to memories, similar to the output that we get to when done consciously with remembering long-term repressed memory recalled consciously, recalling them is sort of similar to recalling a dream. Therefore, one can have access to the memories of the unconscious which is similar to remembering long-term memories. BTM can be thought of as a form of personal hypnosis. BTM allows the mind to imagine using auditory processes similar to hypnotism, in that oneself gives control of processing words consciously to the processing of words unconsciously. This doesn't mean it really happened or it is true, similar to a dream, it is the possibility of something that the selfs, parts of the mind, are investigating and possibly discovering something new. You could say that the hypnotist is the audio track getting played back and the patient is the one listening and doing BTM. One finds the probability of something happening and can relate this to real life, to warn or to motivate, to protect or to find the way to find obtainment of some goal. The state of mind is important for BTM because the state of mind or brain waves are the sieve to collect different answers from BTM. They will come forth depending on one's viewpoints. Someone skeptical may find nothing or a different viewpoint of the BTM than another person. Someone believing in BTM may find only one side of the story. So, while if the observer has a different initial impetus by discerning other possibilities, they will get different brain waves and then be able to perceive the other viewpoints of the story each time they do BTM on the same audio track (*).

How is intuition related to the unconscious?

Note, in the underlined title above there is an SV about: **~[. . .]if your chakras are not in tune, intuition can mess you up.~**

1-5-0 Can one know the subconscious in real time from the sense of intuition? Yes, intuition to me is just a general gut feeling and doesn't separate it from which sense is primarily receiving the information. The intuition could have been triggered from any of our five senses.

The unconscious on the other hand is our programming and is double hidden to any outside probe. The unconscious can only be reprogrammed with change in the person to a new **+self+** if they notice the change from the subconscious. For example, if I want to change my diet to be more vegetarian, then I will pay attention to subconscious ques to eat fruits or vegetables.

I can get an SV message to eat apricots and then when I'm at the grocery store, just ask myself, "What else do I need to buy today?" The answer I got in my head was to eat apricots, "Can you believe that." By listening to SV it made me remember by just asking the question to myself. If you don't notice that you can change, then you are sucked into your current program and none the ever wiser. However, if you find subconscious clues that you are on the wrong path in life then it is possible to peal back these veils of consciousness and commit to change, over time, and reprogram yourself. Therefore, intuition is our natural ability to change by helping to reprogram the unconscious.

How is intuition related to opening the third eye?

1-5-1 One can open the third eye and have new levels of intuition and consciousness. The levels are different types of clairvoyance: for example hearing someone talk to you, clairaudience; seeing symbols in your field of view is clairsight; and feeling someone's aura is clairtouch. Instead of just getting a gut feeling of something happening or about to happen, the third eye can add additional layers of detail to help understand where the basic intuition is coming from.

Definition of My **+Self+**

1-5-2 I believe the thoughts that are broadcasted in our brain and from our mind, without using the throat or tongue, are our conscious thoughts. If one is conscious then they can hear themselves with their silent voice, because this may sound different than the spoken voice with the tongue. The term that I will give this is **+self+** because it is from one's brain thinking, or from a specific thinking person, then this should be capitalized as a noun is, for a particular person's **+Self.+**

The silent voice, I believe, is associated with the unconscious of an individual. The reason why it is associated with the unconscious is because I found **+self,+** **+me+** and **+meself+** during SV analysis. This means to me that while thinking silently a person has direct access to the unconscious and is able to program themselves with their thoughts if they have the tools, skills and the wherewithal to make it so.

The **+meself+** is the same as the higher consciousness. On the other hand, the spoken voice or the sound out the mouth, off the tongue and heard by others I would identify as someone's CV.

Definition of consciousnesses

1-5-3 I believe consciousness is shown by our personae, the many ways we can show on the outside to the world who we are and how we interact. Also the unconscious **+selfs+** are: those thoughts, symbols and words to hear in our mind that are inside; that voice in our head when we think without using our tongue or throat; and the many ways we can show who we are from our insides to outside.

The **+selfs+** are constructed by adding all the thoughts that our brain can have together at a time into reality. Everything from the brain starts with a thought. This thought can lead into a symbol or a voice and we can see or hear internally just in our mind, I identify this with **+self.+** Some thoughts are in our awareness because we can understand the choices it represents, as in an idea, or it is represented by words or pictures, as in a symbol. I can explain this by how someone normally reads a book.

While we read, our eyes see a word and the mind reads it. Meaning, we read it by not speaking or even moving the tongue. We can read it completely within our mind, with our singular conscious focus as the **+self,+** so we can understand the word, and its meaning in the sentence it is used for. One can ask their **+self+** a question telepathically in one's mind and subconsciously know the answer.

At least for me, the volume of the voice that words are pronounced with, in my mind, can not be amplified. The only way to soften the impulse of hearing a word pronounced in my mind is to divert its focus to another thought. Because the thought of speaking a word in the mind brings it into consciousness and we are focused on it. To make it softer, just think of something else, for example divert some of your attention to, say a picture.

I can imagine a word, while also still putting more of my focus onto an image. Of course an image has nothing to say, but things in the image are described by words and symbols. I can start naming off things in the image, but still remember the word I was reading seconds ago, although more diffused, I moved it to the side of my focus.

Here on the side, I have short-term memory working. While the word is in my short-term memory it is dangling in my mind in the vicinity of the focus, in the part of my mind which is presently understanding an image by naming the things in it. Subconsciously, If we forget what we were thinking to remember, then this memory should be going into the brain somewhere, as an unconscious memory.

Hence, since our minds have one main focus, at least one side focus and an unconscious pot of information. If someone has time to remember, then the word someone was just reading is no longer voiced in their mind, it is in the side focus as a symbol. The parts of the image now are going into an area of the brain to consciously remember them. For me, I might want to remember things in the image to tell my friends about.

My friends might ask, "Did you see that picture?" And I would say, "Yes, there was a tree and green grass." One reason I want to remember, is to be sociable and be able to describe my surroundings to others in my life. This gives me the opportunity to keep or start friendships. "Yes that grass needs to be cut," I would say. Then this gives an opportunity for my friend to change the subject, "Yes, and it really rained last night so I will be

cutting my grass soon, probably on the weekend."
Hence, consciousness is the ability to interact with
our five senses to solve problems, remember ideas or
interact in situations within our surroundings.
When this happens with other thoughts that are not
in our awareness, we are witness to something
outside our problems, ideas or situations, then we
enter into our intuition and the unconscious. Here
is my definition of consciousness from before, on my
website.

    1-5   Definition   of   Consciousness:
Consciousness is the result of our
unconscious and subconscious working
together. The unconscious is as the soil for
an onion and the subconscious are then the
roots of the onion to the soil. With BTM the
voice backwards is the subconscious and for
the onion we can see some but most of the
roots are unseen and below the soil. The job
of the consciousness is to make more shells
adding to the onion bulb. With more shells to
the onion then there are more roots and then
more leaves. So the consciousness is
everything on the outside of the onion that
one can see. The more connections we have to
the soil then the more connected one is to
everything in one's soul. Root connections to
the other shells of the onion represent other
parts of ourselves. Each Self has roots, a
shell and leaves connected and growing
together. I believe that our soul wants to
provide the onion bulb to grow to the size
required so that the soul is the same size as
the bulb, full and complete. The onion bulb
has many shells or parts of the person, with
perhaps the personalities as shells, these
selfs usually do not communicate to the
conscious, out right. These selfs inside the
onion as shells of consciousness and allow
the soul to express the essence of a being in
life. So that to have a soul is to be by
default a disconnected consciousness from the
unconscious so that the conscious may live
and grow to follow a path for what one wants
to do in life. The core of the onion, I would
say is as our adolescent mind. As one gets
older more shells to the onion are added and

one develops a personality from the additional selfs that are growing together. The outer shells farther away from the core are the more conscious parts, including personalities for different life situations. Outside the shells of the onion we can see the conscious Self, while under the soil the roots are the subconscious and the soil within the roots is the unconscious. Then the soil and air would be the base and our connection to the unconscious which is our connection to dreams messages and universal collective (*).

Definition: +**Selfs**+ and our Internal Psychic Parts

1-5-4 +**Selfs**+ are, I believe, individuals or at least things, in that they are the parts of one's thoughts and they can take on many forms in the unconscious and conscious. I have identified +**selfs**+ that have language from lower energy to higher, corresponding to: +**ogres**+ or the id, as Sigmund Freud might say it is, which is more primitive yet genetically older than any of the other +**selfs**+; the +**self**+ similar to the ego, which is most often what is heard in the mind when subconsciously thinking and silent with the tongue or the intuition; +**meself,**+ the superego which is the wisdom of the subconscious and then lastly the more unconscious helpers or guides from other dimensions.

I think of the +**ogres**+ as a type of lower-self. After searching duckduckgo.com there are others who have examined this as well as Sigmund Freud on his Id, Ego and Superego. I'm not claiming that his Id, Ego and Superego are one in the same as the +**ogres,**+ +**self**+ and +**meself**+ nor am I claiming that the lower-self is equal to the +**ogres.**+ The **I** as the conscious voice in my head, **I** sits somewhere in within my brain. The **I** is the drive for self-expression while the +**ogres**+ are the drives for self-gratifications.

The +**meself,**+ I believe, is what controls the process of reflection and intuition, to moderate and direct the other more conscious +**selfs,**+ and so the

+meself+ is mostly in the unconscious. The **+self,+** I think for SV, is related to the body. Add this to the others and then we have what I call the total psyche of the unconscious: **+ogres,+** **I,** **+self,+** **+meself+** and guides or helpers.

The conscious **+self,+** I believe, should be labeled **I** and this is what we hear in our head when we are just silent and thinking with our brain and hear our voice without speaking. This inter voice is not what other people can hear but it is the way, I think, we think others will recognize us as we are pretending to speak-out loud. The **I** is communicated in our mind when we choose to focus our attention on thinking. If at the same time we are also typing on a keyboard then these are movements of the body parts and they would then fall into the subconscious.

Something would be unconscious, for example, when the body wants food or drink. The immediate needs of the body are not advertised to the conscious until it is appropriate and sometimes convenient for the conscious because, during the day the conscious takes center stage, the conscious needs to get tasks done that are required for the **+self+** or person to do in any given day.

In conclusion, I would put the total of the **+selfs+** or the person, that communicates with a language, as equal to adding all of the **+self+** identities together, which are: **I** for the conscious, thinking and "I" for speaking; lowest level energy for the **+ogres+** and then higher energy for the **+self+** in the conscious and subconscious related to the ego; for the unconscious, mostly outside our conscious life at the highest energy, and observing and usually not apparently known consciously is the **+meself+** which could have a robotic accent in SV and finally the guides or helpers that are directing our soul so that we may complete what we came here to do on Earth.

Figure 5:

Diagramming the Conscious, Subconscious and the Unconscious.

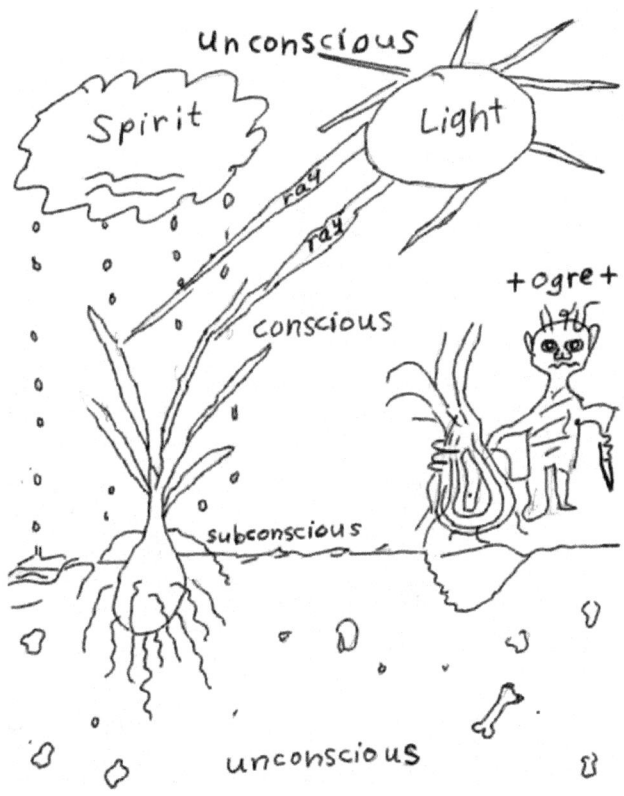

Another of my definitions of consciousness:

1-5-5 Consciousness is our unconscious and conscious working together and connected to everything in our body so to make connections with the whole, as the shells of an "onion." Connections to the other shells of the onion represent other parts of ourselves.

I believe that our soul is the onion and the **+selfs+** are inside the onion as shells of consciousness. The core of the onion, I would term as our consciousness of **+self+** and the five conscious senses. The outer shells farther away from the core are the more subconscious parts, including **+meself+** and guides or helpers. Outside the shells of the onion we venture into the unconscious or roots and into the soil by astral projecting outside the body.

How we describe the conscious doesn't really matter. There are many ways to show by anecdote this subconscious realm, by tying to explain, as there is an intelligence besides the conscious.

>How to relate the dream state with subconscious and unconscious compared to the conscious memory state of short-term and long-term memory (*).
>
>1-7 What one thinks and says has a lot of bearing on who one is and how they express oneself to the world. What we say out loud, the "I," is a reflection on what we have inside, our soul and Selfs. Using BTM we can access those subconscious onion shell levels of our Selfs, for example +Meself+. The +Meself+ is our higher Self while the +Me+ is the talk in the mind that one can consciously hear internally. The lower consciousness onion levels, I think are more of at our core, conscious +Self+ and the senses, including the Ego. The +Meself+ and +Me+, are interacting with our other onion shells in the day, during sunny times or days with rain, moisture in the soil and rays of sunlight during the day provide growth to oneself. This story of the onion is to show that our mind can interact with the environment to comprehend its surroundings and absorb and grow consciously and unconsciously. This is similar to what we do when we read a book. Each word we read combines together, which is what the author

wants the reader to direct their attention to and grow into a new understanding. Every outside influence has the [possibility] ~~possibly~~ to change someone. Lets first describe change (*).

Change

1-7-0 Change is easiest for the human to express themselves with the use of words spoken to oneself in the mind, silent to the outside world that react to our internal **+self,+** or spoken with the tongue with words, that may react to our outside environment after we think with our thoughts.

How we use our intelligence is important and can result in combining two thoughts together to make a third unknown thought, that we can realize by putting it into words. Because words are quick to produce a change for us, taking little effort to realize and voice out loud. What one has to say after thinking for an instant can have an immediate effect. Since we usually do not stop to consciously think what a reaction will turn out to be, "Do we have time to consciously process what we are going to say?" I think not always. Some of the thinking needs to be unconscious so to be timely to some situations. Words can easily be from both the unconscious and conscious choosing the thoughts to describe with words. Those thoughts can potentially have two sources with the big brain or conscious and a little brain or unconscious.

There must, I believe, be a combination of emotion and logic in our spoken words. Because of this duality in the way the brain is constructed by all of the components of the mind combined together, the mind is a superposition of all the waveform and energies in the parts of the brain. Some of those waveform and energies are expressed in physical reactions in the body.

Nowadays, with technology, we can easily record our reaction with the vocal cords or the spoken words and also with listening from the audio on the internet. We can then change our behavior in a faster way if we find the hidden messages that are in the unconscious.

How I perform SV

1-7-1 The computer with an audio editor program installed and connected to speakers with a microphone has made this analysis easy. Here, I will define an SV audio track of normal conservation as recorded with an audio editor, and then perform the reverse effect onto it. I will define what is continuously listened to, in this way, and deciphered by one's brain. Then, "Why Subresonance Voice?" Because, I see myself resonating to an answer, question or the performance of some attribute of some things in the environment. With what I understand of this SV in my brain, I'm reassembling these snippets I remember while still listening to more of the audio voice, then later I can coherently understand what I was listening to.

For SV, I don't actually mean to say I see in my brain an imagination of a blueprint getting taken apart step-by-step, in exploded view. Rather the "taking something apart" (as in mechanics) means that there is an unconscious part to the SV audio that is analyzed in the deeper recesses of the brain, I believe this to be in the cerebellum, from my SBTM/SV research, but my conscious focus is articulating what comes to the top as subconscious and this becomes consciously aware, as I'm still listening. However, the contents of this information are sometimes uninhibited, depending from what part of **+self+** that they are emanating from.

Some SBTM/SV are extremely hostile or dire and so they are deeply personal and hard to explain to others, from my experience in my American Culture. <u>BTM listened to by others may not be easy or scientifically repeated because subconscious words have</u>[. . .]

(Side note: After waking-up on Saturday May 6th, 2023, I recorded that I remembered in my dream and what I wanted to do at work. I was listening to the SV, I don't do BTM anymore, which is recording oneself and also others because with BTM, the others are able to connect or be controlled by bad spirits. This is what I interpreted the SV as, **^Matt don't do**

BTM on anyone anymore, they can take advantage and take control of you.^ This is essentially the SC message while doing SV in a loop play. I stopped the play on the audio editor and tried to find where this SV was, but I could not find it, it was just SC in my brain from outside, probably. Upon finding where this was after multiple plays, I found the spot where I got the SC again. Here is the SV of that interval, **~Ya shammy gammy be aware of of-good-digg-dig be very aware They-you'ill soo-n of the-they-a-me-a-sege-gegi Matt.~** When I reversed and retrieved the forward dialog it was: "Um he gets it for some reason and whose going the wrong way, but it could go the right way, in my direction.")

Note, in the underlined above (bottom of page 70) there is an SV about: **~If you release this book, . . . sacrebleu,. . .easy easy a lot of people cannot absorb this type of information,~** Continuing with the paragraph:

[. . .]meaning not exactly similar to conscious words. Many times if I need to transcribe SBTM/SV to paper, I need to repeat it again and again to get the order of the words correct even though I know the gist of the SBTM/SV story-line. Because BTM uses metaphors and BTM flips sided when converting to written words. One must ask where this information came from?

Note, in the underlined above there is an SV about: **~. . .for me just to let you know. . .suck for you ¡Ay, caramba! Your life is going to suck,~** this is because when transcribing SV, for me, I need to change focus from the SV and then to the conscious words, CV. Humans can only focus primarily, consciously, on one thing at a time. Of course you could try to consciously focus on two things in one second, but which one would you think first? My answer is, as if, you know someone said you can only ride one horse at one time.

Possible sources for information obtained from SV

1-9-0 The answer to this is that we have different **+selfs+** working together. Sometime we repress some and at other times release others. We show different **+selfs+** in different situations: being

with friends, good friends and also with other older
adult friends; talking to our significant other or
to the one married to; when we are confronted in a
fight or flight situation or fantasy talk to
ourselves in our brain, as our different shades of
our **+selfs.+**

Sometimes we don't have to think what to say, it
just comes off the tongue and out the mouth. If
someone was unconstrained with this strategy they
might find themselves in trouble with other people
and have to listen and work with them. "Where does
this information that they spout out come from?" It
comes from source, not necessary from the universe
as I first thought!

Sometimes we can control our chakras and **+selfs+** and
ask questions and get answers, either through
intuition or SV.

A tactic to use with other people is: find a common
interest as friends or when at work, for getting the
job done using teamwork. Then people <u>find things
natural to talk about without getting excited.</u> . . .

Note, in the underlined above there is an SV about:
**~If you ask me bub it's all because of what's
transmitted within you.~** Continuing the paragraph:

. . .<u>The source of information for BTM is</u> then the
conscious thoughts one is immersed into and also the
source, were we can ask questions, for example,
questions about evidence from intuition or other
physic abilities.

Note, in the underlined above there is an SV about:
~Say it might be just coming from source.~ So,
Instead of not immersed, just say SV comes from
source, the cross of the **+parallax.+**

Rules for using BTM (or SBTM/SV for Matt Mandell,
the author)

1-9-1 SBTM/SV, I believe, we should use this with a
neutral strategy that doesn't harm your neighbor.
In this way I wanted to do SV on myself, publicly
for this book. I found out if people are offended
through the BTM process, then their requirements
will be communicated through BTM. Some frankly did
not want to have BTM done to them, then I stopped

because that is the ethical decision. I remember them because they probably kept the same preference every time I accidentally did BTM on them.

To be fair, I will not steal their ideas from their forward speech. Now, I don't do BTM but, I could listen to them in CV and then do SV on my report of what I remembered that they said. This SBTM information can document, for the purpose of the exposing the hidden information that might <u>be scientific or in other ways make sense to me.</u>

Note, in the underlined above there is an SV about: **~This paragraph speaks for it's own, stop being a Nancy,~** I guess it is good enough, no more changes, continuing with the next paragraph.

This is especially true if it also found that the same BTM story-lines are in other people. To do otherwise and frame others publicly with BTM is not ethical. There is much to study in SBTM/SV and prove without disparaging others. The easiest place to start is the difference between objective and subjective.

We need to think with facts and become informed to different choices so that we have a responsibility interacting with each other and then finding subjective areas that may lead to new innovations. OSA can help us to find better solutions because OSA is not based on any single way of finding a solution, but helps look at different ways to solve a problem and if SBTM/SV shows that an OSA is wrong then we don't have to use it. If only part of it was right then we should examine where this good part leads to.

How our ways of doing things in our culture and sub components of society relate to reality and how we communicate with words and money.

> 1-9 Our way of being is foremost in the culture we are living in. So lets equate how we act to what is the description of our culture. Our culture is further broken down, as from the level of society, or class, to our race and ethnicity and analyzed down further to the way we act individually or as in personality types and dispositions including gender. We know that when humans

communicate we use spoken words. We think words mean something. The notion that reality is objective is so because we learned it in our cultural experiences. When we pay our bills and it gets sent and received, then this debit is withdrawn from our bank account, instantly. There is no arguing here, because if someone buys something they have to exchange money that they earned to lawfully purchase something they wanted or needed. This is an objective truth. But life is not so easy because there are differing viewpoints, twists and turns and the subjective (*).

1-10-0 The dream of the candy store and differing viewpoints as an example of how something subjective may never be objective, "How do we know we are sober and awake to see objectivity?" One can dream that they bought something and say to themselves, "Wow, I'm in a candy store and want to buy candy for Easter." Then, when they wake-up they might be in hot sweats because they are on a diet and can't eat what is outside their diet. The dream seemed real, it can even be considered a nightmare for its realism.

During sleep the one sleeping actually believes that they were in a candy store and going to buy some candy. It is so surreal because now, they are probably still dreaming while also becoming awake. Being both awake and asleep is not objective. "What is the truth in someone if not knowing if they are asleep or awake?" I believe that this is subjective to the situation and can be read in different ways. Given permission, I can find out if this person actually <u>did keep their diet. I'm trying to determine if this event of the dream caused the person to break</u> the diet that they wanted to objectively keep or not.

Note, in the underlined above there is an SV about: **~The problem with reverse speech is that it is not specific,. . .reverse speech is my therapy.~** SBTM/SV and BTM may not be specific, but they are **+all-encompassing,+** since they are in the timeless, the 5D.

Ask some questions to someone about their dream and how to relate it to living in reality to help them analyze their dream.

1-10-1 I would ask the person about this nightmare, "What is your first impression, did you buy the chocolate?" The answer depends on if they are still asleep enough to remember more of the dream. Because the person woke-up before the end of the dream. They answer on a split second, "I feel like I did buy the candy." I would say to this that, "Feelings are not objective and only have personal worth." They might respond back, "Well, I just remembered more of the end of my dream, I saw through my eyes in the dream getting dark chocolate kisses and putting them on the scale to get weighed with my hands."

I could counter, "Is that the truth or are you just making it up?" They would respond, "I didn't just make it up because I remember it was like watching a movie." Witnessing with our sight is more powerful and better evidence of something happening than with our words. Symbols are more powerful than words because words blend into symbols as we transition the memory into short-term memory. The five senses seem to allow for differences in opinion and are not always objective. This is because some senses are stronger than others.

Reaction to the dreamer's analysis, "When do we choose to do something, is it our subjectivity leading us to objectivity or just the opposite, do we have control?"

1-10-2 Observing the one dreaming, there are two outcomes for this dream in real life. The dreamer bought a candy bar in waking life or they didn't. The objective question is easy to solve, just find the evidence. Just wait and have the shopping and eating habits kept in a database and then we can chart a graph to see if this person can keep their diet. The harder question is to know how their unconscious can modify the objective reality. "How does this person cope with stress and joy when they eat?" When the person has stress this might be a breaking point to eat the chocolate.

There might be multiple stressors at the same time acting as a conjunction to amplify the break of the diet. It is often said, "Once a person makes a choice it is their choice because they wanted to do it all along." This of course is not implying that there are threats to one's life, possessions or livelihood. However, someone can trick the person into a choice, or something not in their best interest, they are just not presently consciously aware of what is going down. Even on this count, the person or "dreamer" may be choosing just to be helpful to another and tryout the choice to see if it suits them afterwards.

"Who is making the choice, the dream or the dreamer?"

1-10-3 In the case of helping others, the choice may not have many needs of the dreamer's met. <u>The dreamer just agreed to what was presented to them and often regret their choice because afterward they know</u>[. . .]

Note, in the underlined above there is an SV about: **~This is the kind of paragraph you get if you self publish, just wasting your own energy,~** but I'm including it to show the process in learning SV. Continuing on with the paragraph:

[. . .]they were beguiled into the choice. Their goal of one thing, a want or need, being met was diverted into helping someone's needs, the one with more experience in this situation. So weather the choice is a trick or done in fear, the dreamer was using something in the **+self+** to guide them, sort of being on auto pilot.

When the person makes the choice, actually in their best interest, then it may seem that everything is working out. "But is it?" The dreamer may be reading many ques and following the right path, they tell themselves, in **+self+** discovery without even knowing it. Because the actions from the choices are part of a dreamer's programming. If the dreamer was more logical, if it makes common sense, then the one at the disadvantage may succeed in obtaining their wants or needs from the environment the dreamer enters into, and not the programmer. "What if these trips, snags, harpoons, nooses, fishing, whaling, snitches or snares can be exposed?" Some people would be offended, by those who are presenting these insights. The dreamer could be helping others and not themselves.

The options our dreamer has, "Are they tricked or has fear made them dream or choose some specific way?"

1-10-4 The first two scenarios with the trick or fear, these may make the person feel as though they are going into a downward spiral, after realizing this, as the unconscious was reveled to them. These two, the trick or fear, makes the dreamer the dupe. The last scenario may surprise the person if they just guess the correct path, they are lucky. In this case, they stepped outside their box in life and made a random choice concerning their path, and won. <u>This then resulted in keeping their diet.</u>
. . .

Note, in the underline above there is an SV about: **~Matt Mandell, if you print this book I see a lawsuit, massive,~** this probably refers to continuing what I wanted to do with my book, but things can change. Continuing on with the paragraph:

. . .The best case would then be the person knowing what their unconscious body needs are, and they are able to provide to it with ease. They don't spend time consciously concerning the lesser scenarios. They would probably think the unconscious path alternatives, trick or fear scenarios, to be humorous. The humor is from the consciousness realizing that there really is an unconscious one and then a tragedy appears. "What would happen if we gave a helping hand to the trick or fear examples?" Let me show with an analogy.

Example of the four consumers, (1), (2), (3) or (4) and describing the different outcomes of their choices

1-10-5 There is a dog pen and only those willing to do business go into the pen. Because once the consumer enters individually, then the guards outside will close the gate shut. The consumer, (1), (2), (3) or (4) find themselves in an enclosed cage, with metal fences stopping any way out. They are directed to some dogs in a wooden dog house (the salesmen), because there is no way out. There may be more dogs inside the dog house as the consumers can't see or tell, right now. The consumers, by virtue of walking in have brought meat with them, I hope. Because the dogs are hungry, they are salivating. So, lets get on with it.

Each consumer has a need or want. For the trick scenario (1) and also the fear (2), the consumer brought no meat for the dog's lunch. The (1) got tricked in because he or she thought the cage was a dog shelter. They wanted a puppy. Wrong. I don't need to describe what happens to them next. Suffice it to say, we now know how many dogs were in the dog house. The one who has fear, (2) because they were totally unprepared for the situation, they had a fight or flight response. They tried to climb to the top of the cage to safety if they want to escape, or to try to attack one of the dogs to show who is stronger. This results in the end, not well also. For the one who was lucky, scenario (3), to bring in meat or the one who logically planned out the scenario with SV for the solution (4) has the meat and can wait, for when the guards will automatically open the door for them to get out safely.

There is tragedy or glory as the possible results. If we gave a helping hand to the one tricked or in fear, you could ask, "Would they learn?" For example, scenario (3), the consumer just brings meat by chance, not tapping into the unconscious. If just consciously, probably not, they would never guess to bring meat, they got a "Darwin Award," because they cannot repeat this again. However, if they had access to their subconscious then they might have a chance to succeed, by learning from this.

The choices of these four customer scenarios can be used for psychic experiments in general.

1-10-6 So instead of just using objectivity or looking at the data of what dog got which consumer we can examine the subconscious and look at it subjectivity. Because objective data that can be collected to find out before or predict the outcome is wholly objective, it follows a formula. This can be compared to a psychic experiment, given inputs, "Tell me the path to take so I know the end point location," and get subjective data. This can be done on intuition and not objectively, just start walking and find the end point to the treasure.

It's not always that easy to follow the dotted line and find the place that marks the spot with an X-shape, because of uncertainty, but maybe because of a conflict inside, so let's delve into the subconscious and unconscious to frame reality as a different aspect of psychic inputs and outputs.

"Are we hurting our chances by using too much of our consciousness?" I think today we are, "Why not ask for help from the unconscious **+meself?+**" The answer is, "We really need to know who we are and to try and improve ourselves and see how others see us, then in this way we will see our inadequacies and try to learn unconsciously instead of the usual consciously thinking to improve ourselves."

> Example of the subconscious and unconscious interaction (*).
>
> 1-10 The recall of a dream is one way for accessing one's unconscious. There is a paradox here: "How can we as humans remember our dreams if they are unconscious?" After all, when we are just starting to awake we are therefore not completely unconscious. I would start to answer this question by saying, "Since the only other word to describe the absence or partial absence of consciousness is the subconscious, then lets use this." Then I would ask, "But, how to define this subconscious?" Let me first define that the unconscious would be, from my understanding, that this is when there is no

conscious inputs for actions, the person is in delta waves or asleep. The subconscious state, I would answer to me, "The subconscious describes when in a wakeful state but also experiencing not just the ordinary senses of sight, smell, touch, hearing and taste. I would also include here ordinary sense of oneself thinking." Thinking is that voice we know as +Me+, that which talks silently, only heard internally by one's mind, when he or she thinks. I can [imagine] ~~imaging~~ this sense emanating within the brain. I believe that we should associate this sense with the Ego sense, in honor of Sigmund Freud. The Ego is not wholly the +Me+ but can meld with it. So we have six senses consciously when we want to think subconscious. Since the Ego is not usually consciously thought of, because if you are, then I would consider that you are day dreaming, then I would not include it with the five senses. The soul may be the source of the Ego sense emanating from the realm of the unconscious. The subconscious can be accessed by using BTM and can detect the Ego. The subconscious communicates with memories popping into consciousness[. . .](*).

However, there is another sense to pull us into the unconscious. That sense is the realm of the subconscious, and it can answer the accessing problem in this paragraph above, about dreams. Continuing on with the paragraph:

[. . .]These memories can be represented as symbols and these symbols can be remembered in the mind with metaphor, something that can be objective and subjective. The Ego may not always have the best intention in mind, so one needs to have a method to know when the Ego can attack. For example the metaphor +cinamin+ can objectively mean a spice, cinnamon, to put on toast for the taste or be a metaphor to describe a small sin or sin-an-min. The Ego thinks it can get away with murder by making a small error and then blame it onto the "I." That way the Ego can insert itself into the "I" and the "I" might not realize it (*).

How the unconscious memory is efficiently formed into metaphor and story-line (*).

1-11 The subconscious can be described metaphorically as the +whirlwind+.[. . .]The +whirlwind+ can be accessed[. . .](*).

Note, in the underline above is an SV about: ~**see the subconscious as a robot,**~ in that the subconscious is scary to us and unfamiliar, so it is foreign and moves by unknown engineering.

The **+whirlwind+** can be accessed to examine the Dream, of eating chocolate, by recording dialog from the dreamer when they are awake and then listening to SV. If one talked specifically about the dream instead of writing it down and recorded the audio then the SV can give information that acts as a new sense to the brain. From my personal experience listening to audio played by way of SV, right after a dream, this is a way for me to access the subconscious and document it.

Note: In the last underlined sentence above there is an SV about: ~**Hey you fixed it son, now this paragraph reads a lot easier since you fixed it,**~ so I edited it and this paragraph reads better. Continuing on with the paragraph:

[. . .]by recording dialog from the dreamer when they are awakened and then listening to it play backwards with BTM. If one talked specifically about the dream instead of writing it down, recorded the audio, then the backward playback can give information that acts as a new sense to the brain. From my personal experience listening to audio played backwards right after a dream, a backward track, this is a way to access the subconscious and document it. If I do not remember the dream because I did not record it soon after I got awake then I might just remember a symbol in a dream, for example as in I was in a house. If I would of recorded the dream when I first consciously thought of the dream then I have more information. Going over the dream consciously the first time, then one can learn to remember many details.

So instead of the house symbol, I have more information, because I recorded it, that I was in the house's basement and the roots broke open into the basement concrete slab so water entered the basement. I have not just a symbol of a house, now I have a line of reasoning for why I was in the house because there is a story-line that describes what happened. Just as a metaphor can have a meaning beyond the objective word so can the story-line represent a side story of what is going on in my conscious life. The interpretation is, I need to make a foundation so that water cannot get in then I might be successful at what I'm doing in my waking hours. The answer came from a dream's story-line submeaning (*).

An example of why to use SBTM/SV after realizing how to use this subconscious muscle.

1-11-0 When I first seriously started to do BTM I would listen to company conference calls on my computer for stocks I was interested in. They are usually an hour long and when I listened and did the BTM I would find myself falling asleep before I got to the end. One of the side effects afterward is having a nagging feeling in my brain. I consider this a remnant of BTM where <u>the brain has to modify synapse connections to decompress neural tensions from listening to BTM.</u> . . .

Note, in the underlined above there is an SV about:
~What's near to me it's probably old fallacy,. . .~

Note: Maybe the brain doesn't need a rest during the day, it is just a sacrament for the conscious to take this break. Where the conscious needs a break so that the unconscious can think on its own. This could also mean that the information is opposite from the **+parallax,+** in truth, if it is **~near.~** Continuing on with the paragraph:

. . . Eventually, this nagging feeling would go away or get relieved. This is similar to how we feel refreshed after having a good night sleep with a dream. After a while there was harmony by taking a two day break from BTM on the weekend. <u>This allows for a reset in the brain so</u> . . .

Note, in the underlined above there is an SV about: **~Matt it is hard to explain this I do not have all the answers to explain this, I don't have the answers here.~** Continuing on with my comment paragraph:

. . .I can, I would say, better focus on SBTM/SV. Because SBTM/SV is laborious in that I can only listen to so much every day, as I don't speed-up the SBTM/SV, I keep it at normal speed. However, the more SBTM/SV that you do then the easier it is accomplished. This is a telling fact of the brain, in that it gets stronger with repetition. This muscle strength is similar to the skills used to solving a puzzle.

> The connection between the pineal gland and using BTM (*).
>
> 1-12 Once the BTM puzzle was first solved, for me, there was an epiphany or little tic-tack that went snap in my brain. When I think back, it was located suspiciously in the area where the pineal gland is located. After this, I eventually wanted to eat more vegan in order <u>to decalicify my pineal gland. Because this gland is in</u>[. . .](*).

Note, in the underlined above there is an SV about: **~Sit down and breath out will help you develop psychically.~** That makes sense since people do this in yoga class and doing breathing exercises is a way to help with psychic awareness. Continuing on with the paragraph:

> [. . .]the brain as a singular gland which seems to allow for two sides to come together metaphorically and physiologically. When I opened my third eye, a little, there was a new sense of reality. This new sense in my opinion allows me to more thoroughly allow myself to understand the different scenarios

evolved while listening again to the same BTM and see it in life.

How the five senses are prioritized to solve a problem quickly. (*).

1-13 One's choice of written words to describe a painting, for example, may be different from another person's written description. <u>While on the other hand given ten seconds to look at a painting, one can only remember so much.</u>[. . .](*).

Note, in the underlined above there is an SV about: **~I bet it has to do something with reverse speech is magnetic, you don't hear that enough love.~** Wow, who came through there? Continuing on with the paragraph:

[. . .]The personal account of one person to another can still vary given that they may focus on different aspects of the painting, in the allotted ten seconds than another observer's ten seconds. There are probably psychological tests that can test the relative strength of one's senses to the public at large. While writing this book, my first impression is that everyone's senses are at the same resolution and strength. Upon thinking, this can not logically be the case. For example, our taste of wine may not capture the flavors present to a wine connoisseur. We all remember the experiences when we are focused on solving our favorite problem. When solving that problem we don't pay attention to the other senses, this might help us solve the problem quicker. Probably because we sense urgency and this turns some senses off to divert attention to the senses that maximizes effort to complete the tasks in solving the problem, probably just the one dominant sense in use. An experiment, as above, with the time limit can help document how well different people's senses are used and can be compared to what they observe and the strengths of those observations compared to the senses from others in the group. (*).

Example of how to do prioritizing [through] the senses. (*).

1-14 If we are studying a painting for ten seconds then we are not paying attention to what we are tasting. Our taste sense is probably neutral and this allows more brain power to be applied to seeing and remembering the painting. We know the mind is plastic and sets of neurons can turn on and off. <u>This on and off I believe helps keep us to be conscious. Consciousness is the use and[. . .]</u>(*).

Note, in the underlined above there is an SV about: **~Matt sue you, say this is a hunch, svadhistana is the weakest one and the anja is quite the opposite.~** On July 7, 2023 I found, **~Matt, that chakra can sin if you keep it open Matthew.~** This is referring to the svadhistana, I think, if one can't control their thoughts. Continuing on with the paragraph:

[. . .]diversion of resources in the brain, by neural switches, when we are looking at the painting, in this example, while the other senses are now more subconscious because the focus needs to be on sight, to see the painting. I believe consciousness uses the combination of senses in our whole body, including the chakras. The senses includes from inside the body to outside: the five senses inside; for the subconscious which includes the +Ogres+, Ego and +Meself+ and access to the +whirlwind+ and finally the unconscious as the seventh sense. This adds up to seven senses and there are also seven chakras. The chakras, I believe, emanate as a field outside the body and are related to the senses which are primarily inside the body (*).

How our senses can be thought of associated to our chakras (*).

1-15 First let's list the senses: sight [one] ~~(1)~~, taste [two] ~~(2)~~, touch [three] ~~(3)~~, hearing [four] ~~(4)~~ and smell [five] ~~(5)~~; also including the subconscious [six] ~~(6)~~ and unconscious [seven] ~~(7)~~. Next, let's list the parts of the psyche: +Ogres+ [below] ~~(A)~~; +Self+ or +Me+ [middle] ~~(B)~~; +Meself+ or +whirlwind+ [top] ~~(C)~~ and guides or helpers [above] ~~(D)~~. Using these two sets I will match them to the seven chakras, for where I think they go. Starting from the Earth and moving up we first encounter the muladhara [1st] ~~(one)~~ or root which corresponds to the +Ogres+ [below] ~~(A)~~ and smell [five] ~~(5)~~. Next, we move up into the svadhistana [2nd] ~~(two)~~ or sacral, and I believe this is part smell [five] ~~(5)~~ and also taste [two] ~~(2)~~ and also +Ogres+ [below] ~~(A)~~. <u>Next, move up to the manipura [3rd] ~~(three)~~ or solar plexus which corresponds to our touch [three] ~~(3)~~ and[. . .]</u> (*).

Note, in the underlined above there is an SV about: **~. . .if your chakras are not in tune intuition can mess you up.~** Your chakras can be not working when you are under psychic attack. I had a dream that was not logical. Later in the morning when I was up I felt a load on my shoulders that was hard to explain where the negative energy was coming from. **~The problem with psychic attacks is you don't know what you are saying.~** When doing SV listen for +sheep server+ or +caldera+ to give an indication of your plight. These are bad and so be careful of your choice of words. Continuing on with the paragraph:

[. . .]some part of the +Me+ [middle] ~~(B)~~. The next chakra higher is the anahata [4th] ~~(four)~~ or heart and hearing [four] ~~(4)~~ which corresponds to the +Me+ or Ego [middle] ~~(B)~~. Then next, to the visuddha [5th] ~~(five)~~ or throat which corresponds to hearing [four] ~~(4)~~ and [middle] ~~(B)~~ +Me+. I explain this, that one needs to hear in order to speak with the throat, so these two are somehow connected. After hearing we move into inner sight with the third eye and subconscious [six] ~~(6)~~ or pineal gland and the anja [6th] ~~(six)~~ associated with the +Meself+ [top] ~~(C)~~. Finally, the sahasrara [7th] ~~(seventh)~~ or crown chakra corresponds to the unconscious [seven] ~~(7)~~ and guides or helpers [above]

(D). With the unconscious, it surrounds all the chakras and is also connected to the universe, it is the outside connection for us that is connected to the unconscious. Asking for permissions from one's selfs can help direct what needs to be fulfilled for one's life. For example, don't fight against the +Ogres+ or +Meself+, try to find where the harmony is or could be (*).

How the chakras can be connected to the Chi or our life force (*).

1-16 Our chakras are similar to our Chi or the life force in a living body. After reading basic information on the internet about Chi, I understand it to be our connection to the universe much similar to how the crown chakra is the connection to the universe for yoga.[. . .]Using an OSA here, if the bone is to our body then the tendon is to our chakras as the universe would be the muscle. Thereby, informing us that the universe can leverage our lives to do certain callings in our lifetimes so as to fulfill the expansion of the universe. I mean, <u>to connect this to the expansion of the universe because progress, as the expansion and diffusion of the universe, can't happen without everything inside the universe also learning and progressing</u> throughout life also. Version: 41322 (*).

Note, in the underlined above there is an SV about: **~Sac Robinho it's in the sauce that you are asking me but that lateral it might actually be energy from the shark attack, cyclops I'll tell you a dirty little secret that it's a canopy.~**

Note: A search of duckduckgo.com gives bag in French for **+sac+** and **~Robinho~** was of Brazilian soccer, played forward. Probably something here about passing the soccer ball between the canopy above our heads and our own biases "bag" the forward so watch out for the side pass, as to what light comes from above, from the universe that can get through our canopy. The "forward, could also be the ego, and to put the two together then we need to "bag" the ego.

For ~**in the sauce**,~ that means to me, that I'm connected to some intelligence that is supplying me with some information, but can't get all the bits. I have not had an SV as nice as this for a few days! Also, a second order SV here at the underline, ~<u>**He she that uses the subconscious, that is the +gossamer,+ its function is to shield the aura, that definitely needs a rest on Sunday, if you don't do this there is a caldera.**</u>~ Wow. Doing SV evidently creates a thin screen or **+gossamer+** that allows some light through as a filter similar to suntan lotion.

So, SV is filtering an emanation, from the "somewhere," that is not seen, and I don't think this is light. This emanation would be invisible and this emanation may be connected to the ~whirlwind~ because I believe that this is how we instantaneously communicate through the hidden dimensions. Upon further SV, I believe that when doing SV, some of myself is in the astral plane and therefore the **+gossamer+** may make the one doing SV as a ghost in this astral plane.

When I think of the next higher set of dimensions, I think of 5D, this is five dimensions organized together as one. The 5D can't be separated to other dimensions, the dimensions in 5D are stuck together, just as the R3DT three dimensions are stuck together and associated with time. Since time is used in the R3DT, there is no time component to the 5D.

I believe the 5D is associated with a **+whirlwind+** in two dimensions and this is coupled to a new 3D. This is similar to our R3DT realm that is coupled to time. However, the 5D is somewhere above us where, "[. . .]in Earth as *it is* in heaven." above, then so the 5D is in heaven (Mat. VI:10 ABS, 1858). While the R3DT is on Earth, then the next 3D with the 2D **+whirlwind+** is the 5D. These two similar 3Ds in different levels are primary different because of time in R3DT and frequency in 5D. There is a wind that can hit our spirit's body and that is the reason we have a **+gossamer+** to protect us.

Chapter 2
Our two brains acting together (*).

In this chapter I use OSA to argue that my GUT is true or on the right track. I want to go into the process of finding hidden knowledge. I describe the use of Table 1 for documenting the SV process. This might be used for someone else to try to duplicate my results and even use this Table 1 in a scientific experiment. This can be done if one can believe in themselves.

> 2-0 OSA is a way to fill in the blanks as you connect the dots. What is happening in the brain is that the OSA thinking process is bootstrapping a thought and giving credence so, that it can be possibly true, to give some idea the benefit of the doubt, to see where it leads logically and as possible and lead to an end point. That is how BTM works similar to the way one thinks in deriving an OSA. For the OSA or BTM one is using two brains at once. These two are the cerebral cortex and also the cerebellum. The cerebral cortex is consciousness while the cerebellum is unconscious. The point they meet at is in the center which is the location of the pineal gland. While thoughts are crisscrossing there, they can combine information to realize the subconscious in a conscious way. This consciousness is the feeling of awareness associated with the information that is connected for one to think and connect memories to and realizing any implications from them. An example of an OSA is the origin of the universe OSA. In this example, I try to use OSA to understand how possibly the universe works as in a GUT. With the theory of the GUT there is reason to find confidence that BTM information is real and can be useful because there is a theory behind it. As a side note, when studying the functions of the brain with BTM I found that ~the diencephalon communicates with the cerebellum to play hacky sack~ (*).

Example of an OSA on the GUT (*).

2-1 The beginning of an OSA has an objective input to the argument. An objective statement is, Einstein's mass to energy equivalency formula is E=MCC. Energy equals the mass times the speed of light times the speed of light again. This is the first step in the process of making the argument for the end result, a GUT of the universe. Next for the second step I will state that the universe, I believe, has a constant mass because the speed of light is constant. For this second step, I'm building from E=MCC to something not absolutely objective, this has to have some of my belief put into it. Someone may argue that the universe may be still expanding and therefore adding mass to itself. But, I'm guessing that this is not the case. For the last step one must gather one's experience for plausible further explanations and try to link step to another idea or OSA so your theory is not contradicting itself (*).

Another argument for the OSA on the GUT (*).

2-2 If the mass of the universe is constant there might be a GUT of the universe to describe its nature. I know of the big bang, but lets have the big bang happen continuously as an oscillation or rotation, see Figure 1. Repeating the big bang over and over, with the universe expanding to a time then collapsing and then expand back to the time and a little bit later in time. The cause of time might be the expansion of one or more dimensions. This is sort of similar to a transistor feedback loop with oscillations afterwards that are increasing. This <u>can explain how time advances</u>[. . .] (*).

Here, in the underline above, I found using SV the information of: ~**This is an example of an opposite.**~ Continuing on with the paragraph:

[. . .]for an electron, but also going backwards for a positron. [It's] ~~Its~~ possible for information to go back into time while we

90

can physically only go forward in time. There is also evidence of the quantum sea of uncertainty at the Planck distance of magnification. I'm putting this together as a painting, adding strokes or evidence to what makes sense. Similar to how we know from an OSA that a table needs at least three legs (*).

How the universe is divided up (*).

2-3 Continuing with my universe OSA example, space would have to expand and then contract. How this might be possible is through the String Theory, I know in one of its solutions, there is an eleven dimensional universe. I live in three and there can be three higher dimensions not seen by me as well as three anti dimensions. The three higher and three anti dimensions might be disguised by a two dimensional area or shape around our real three dimensions that is blocking the other two sets of three dimensions. Two is a Fibonacci number and is also a magic number. Two dimensions makes a plane or surface. Three sets of three make nine dimensions or three squared, 3*3, or three raised to the second power. If there are two dimensions connecting the two sets of three dimensions such as a worm hole that is not seen by anyone then this would total eleven dimensions. These would include the real three dimensions, the anti three dimensions, the higher three dimensions and the two dimensional worm hole. The worm hole would be just as walking through a mirror to go connect the higher and real dimensions together, "Why?" Because to connect the higher to real are to show how the universe contracts and expands (*).

How the universe may trade strengths between realities as there are a set of fundamental inputs that are varied systematically with one always central and dominant while the others are less dominant and they all trade roles in the other realities or sets of dimensions (*).

2-4 The contraction of the R3DT results in the universe contracting to the source or zero dimension. <u>Then there is a process for the</u>[. . .](*).

Here, in the underline above, I found using SV the information of: ~**There role is the opposite.**~ Meaning, that there are opposite forces in the universe that essentially creates the source out of nothingness and then back again. Continuing on with the paragraph:

[. . .]dimensions to expand to the R3DT. The R3DT are bypassed because this is a circular circuit. The <u>flow would proceed if we take source</u>[. . .]

I found here when I was proof reading that the underline is: ~**Source gives you faith, this stop the wolf.**~ Even though "source" is sounding the same in reverse as forwards—a phonetic palindrome, I included it here because this SV came out clear and it has the **+wolf+** included in a story-line which I have not heard very often. I also included this because right after, in the SV, ~**Matt please take your time to translate it.**~ ~**The reason why you are hearing a lot of messages is because the spirits wanted to talk to you, Matt.**~ Continuing on with the paragraph:

[. . .]as the reference: source with big bang to higher; higher with anti dimensions to real; real with two dimensional worm hole back to source. In the higher or anti sets of dimensions, time would not be continuous because it has to quickly return to the center R3DT. So just as in the R3DT have continuous time the trade off for us is that we do not have continuous knowledge of the position and momentum for a particle. The subjective insight here is that the anti dimension would know position but have no continuous momentum or time. While the higher dimension would have knowledge of momentum but no continuous position or time. It is as if time is changed into a frequency or the reciprocal of time for the higher and anti dimensions. The reciprocal of time is a frequency. This makes me think that the three sets of three dimensions are connected and are balanced together by their strengths and weaknesses. Source would connect the higher and R3DT and the anti three dimensions would allow the higher dimensions to contract back to the real by use of a complex two dimensional shape or shell. The connection of this two dimensional shape that allows the boundary between the real three dimensions separates from the source and the higher three dimensions (*).

An obvious mistake in describing the universe (*).

2-5 The anti three dimensions would know position as we know time. It would be getting smaller and smaller to the source because change is not measured with time but with decreasing dimensions. Sort of similar to going to hell everything would get smaller. Likewise, the higher dimension starting from source would have momentum experienced as we experience time. Momentum would get faster and faster until it reaches the speed of light <u>and then it enters the R3DT space where time is continuous.</u>[. . .](*).

Here, in the underline above, I found using SV the information of: **~That is not the way the universe is do a push-up, you are full of it and certainly a madman.~**

Note: I guess momentum is not proportional to time, I will not use this in my OSA for the GUT. Continuing on with the paragraph:

> [. . .]Upon this transition the weight of the universe is transferred from momentum to the universe which leaves a massless particles of light, our photon we know. The photon can be thought of as a messenger force and not just one particle. Because what makes up all of the dimensions together is nether one zero dimensional source nor a solid but circular movement of energy and matter around these dimensional connections. The transfer of mass

of the universe over such a small distance,
the Planck distance, at the speed of light is
a Planck momentum. The process of the
universe vibrating is a process of
circulation into, through and back again.
Dark matter then would be the mass that is on
balance in the processes of or in transit
within the anti dimensions <u>and the higher
dimensions.</u>[. . .](*).

Note, originally, in the underlined text, in my
manuscript I had the word, "source," included, but
deleted it for some reason, **~Something you got
right, +source.+~**

Note: Ah, so there is a real concept of **+source+** as
in a source to the universe and probably getting
light language from source. Continuing on with the
paragraph:

[. . .]In conclusion of this OSA, I'm not
sure if this is even one percent correct, but
it is the sort of thinking used to find the
story-line for BTM. The story-line is similar
to OSA, where the input is a brief piece of
information realized when doing BTM. When I
do BTM again, with the same audio, I will
hear the same thing and also gather other BTM
information from before or after it to form
more of a story-line. Instead of just
conducting a thought experiment one can
gather information by recording and checking
with BTM and build onto the story-line (*).

Describing the +whirlwind+ and the conclusion
that there is a 2-dimensional boundary in our
[U]~~u~~niverse (*).

2-6 I watch [a] lot of videos in the internet
that describe the fifth dimension as an
astral plane. This might be correlated to my
example of the universe, that I thought
through, with OSA and the two dimensional
boundary. This is how we leave our body in
the state of sleep and can go into different
dimensions. The +whirlwind+ I believe is a
result of the changing of dimension described
in my OSA of the GUT. Having this working
hypothesis helps me to believe and use BTM

> because it doesn't seem as mysterious without this understanding. I think the +whirlwind+ works with the astral plane. That two dimensional area that we can enter through when we are in deep sleep. <u>Since time is switched to a frequency the +whirlwind+ and astral[. . .]</u>(*).

Here, in the underline above, I found using SV the information of: **~Surf side the whirlwind is just a reflection that is why it is the opposite.~**

Note: **^the whirlwind is located in the crown chakra^** and the iceberg metaphor is evident here with the **+surf side,+** indicating near the shore of the iceberg where there can be waves, that whirl or crash down because the top is going faster then the bottom of the wave, and the wave rotates near the shore because waves near the shore become concave and seem to rotate when they crash down and up the beach. "I'm curious what makes the wave?" **~What makes the wave is psychic energies from the chakras, Matt.~**

Perhaps as we know the wedge or surface under the water becomes quickly shallow, allowing the energy of the surf to be multiplied into the wave as it becomes more shallow. As I know that BTM/SBTM/SV can give just the opposite information, especially in market trading as price discovery can move similar to a snake trying to find support and then goes the opposite way to find another support before going back above the previous support. Continuing on with the paragraph:

> [. . .]plane can only be realized in an altered or subconscious state, for example dreams in the unconscious or doing BTM in the subconscious, because they by nature, are timeless. Dreams are but one altered state, listening to BTM is another form of altered state of consciousnesses. OSA is also useful in understanding and deciphering dream meanings. I believe that accessing the +whirlwind+ can be done in more than one way. BTM is one way that this works, because I get the story-line ~~and can~~ from this [to] have understandings in my life (*).

That story-line suggests that I should not listen to some BTM because I will go into this person's **+whirlwind.+** SV is another way to access one's spirit, don't do BTM, I just SV and SBTM now because of the trouble of going into someone's spirit.

> 2-7 The +whirlwind+, I feel can be, felt intuitively without BTM just when you first notice someone and you want to stare at them. I don't want to stare too long else they might notice that I'm looking at them. For the brief time I am staring I feel I can pick-up on something and that something might be the +whirlwind+. If this is true then our intuition or <u>first impression of someone is just reading something of their +whirlwind+. I wonder how the +whirlwind+ is related to the soul. The soul can be damaged and the soul can be repaired in the same lifetime.</u> [. . .](*).

Here, in the underline above, I found using SV the information of: **~Simba. About five seconds is the limit—very creepy. Me posting it, for sure then me see it costly as a lawsuit if you if you stare at someone else's whirlwind, be aware of. I would then be infamous maybe break the sacrament and *sacrebleu* for sure.~**

Note: That's what I did yesterday, but, I was less than five seconds. For **+me posting it+** means putting something on their **+whirlwind,+** I guess whatever I was thinking. This is a reason to redirect myself to good thoughts if they go astray and don't stare at people, at least not more than five seconds. Continuing on with the paragraph:

> [. . .]<u>How it gets damaged is by</u>[. . .](*).

Here, in the underline above, I found using SV the information of: **~The opposite. . .sho-me. . .take you out.~**

Note: If this is ~**show me**~ then perhaps doing the opposite of what your soul wants to do will take you out of life, a bummer. This might be why BTM/SBTM/SV is taboo, because it can divert attention away from someone, that has a good life, and make them loose focus on what there soul's mission is doing in their life. Continuing on with the paragraph:

> [. . .]<u>making the wrong choice and not believing in the possibility that one's soul wants to progress</u> but can't because they refuse to become enlightened and progress in a new experience to evolve their soul. Version: 41322 (*).

Here, in the underline above, I found using SV the information of: **~Sluggish server if it snowballs so you don't stop and then you will find that your star will win the game.~**

Note: Once you get your life going then you don't have to worry about what life you are leading. Let's conceive of it from mind to reality. **~The reason why you fail is because you can't conceive, Matthew.~** You don't have to do SBTM/SV if your life is on track. The **+server,+** I believe is from this SV and is your **+meself,+** a player in the subconscious game of experience.

2-7-0 Continue to listen and even record yourself, especially for describing your dreams. I have created a Table 1 with general questions that can be answered with SBTM/SV.

My Table 1, at page 6, has on the columns the: Why, How, Where and Because. Sometimes more than one can be identified in the SBTM/SV. Above this on the rows are, included also, the modifiers to help answer: Who, What or When questions for Revelation, Collective, Agenda and Description. When I got a BTM for someone who requested not to be BTM then I wrote "Don't do BTM", in the corner or did not even record them.

If there is a match between the two axes, vertical and horizontal, then I write a number where the vertical and horizontal crosses together in the box. I or someone else can verify what they heard and also write details in the notes at the bottom corresponding to the number in the box. This can include the time along the audio track playing in reverse and information about what I call the story-line.

How to use Table 1

2-7-1 Table 1 is a way to document story-lines when doing SV. Sometimes while doing SV, I wonder how relevant something is to the CV. For me, Table 1 deals entirely with SBTM/SV and tries to separate the Who and What questions from the other qualities of a thought or event. The other questions to consider when forming a conclusion are: Why, How, Where and sometimes Because; to expand on the details for an event that happened.

For example, if there is a mass event (more than one data point) in the SV, then I could circle "C" for Collective and if the SV has no information on that box for the horizontal axis descriptor then draw a horizontal line in the boxes across. So if the Collective event has no Location information for SV, then there will be a vertical line in the Location column concerning this Collective. In this way, someone can try to answer questions of the truth and veracity of the information in the SBTM/SV.

For whomever is documenting this, the next step is to record the time elapsed on the track in a box with other notes. Finally, try to find the story-line from another source, then add this information too.

2-7-1 I have recorded an audio song that has reverse back-masking in it. I identify the reverse back-masking stylistically with the reversed virgule, \ \, and the lyrics in quotes " ". Making some OSA quotes and adding a rap to it, with the reverse back-masking.

There are raps about doing the same thing again and expecting different results. "Johnny can't expect anything new" \Johnny is a good guy that can change for the better,\ "because he does the same results, you knew" \Johnny is making wise choices that are not repeats of bad choices.\ "Break those bad habits Johnny because you are getting hit bad" \ avoid the hits and be calm to see the way out don't be a prisoner Johnny.\ "Listen to those new leaves and leav'em and start a knew one" \grow new growth in fertile damp nutrient rich soil, the better for your future.\ "A wise man once said that the tree of knowledge is also the tree of life, think about it Johnny" \go to the tree that helps, just not the tree that someone told you, do the right thing Johnny.\ "Eventually, the unconscious can bleed out and change one's biases" \change to the biases that help, not hurt.\ "Meaning that Johnny apple seed is making some wise trees now. Because he wants to change and be the Joey not that guy, if you know what I mean."

There are many ways to access the subconscious: some may be good, while others harmful and others from out of this world.

Chapter 3
List of ways to access the subconscious or unconscious: (*).

3-0 Listening to music that you love can illicit strong emotions in one's psyche. The music releases stored memories that find their way back to the conscious. I don't know where they are stored in the brain, but they are there along with one or [more] ~~many~~ synapse networks. These memories are not readily accessed because they are repressed away in the unconscious. Their recall is implemented by the thoughts posed in one's mind because there is an association that triggers a set of neurons that do not normally get consciously exposed from normal thoughts in one's day-to-day life because the thought triggers are not there (*).

3-1 Taking magic mushrooms that release the chemical DMT, or N,N-Dimethyltryptamine. The psychoactive drug that is in the pineal gland in very small quantities. This drug can permanently scar the brain and essentially turn it into a heater. All of the important connections are lost and the brain is essentially fried at the synapse. One symptom of having the third eye open too much is hitting or even banging one's head against hard objects. This can also cause brain damage (*).

3-2 Recording your dreams once one awakens. This can train the brain to realize the messages in the imagery of the dream. It has been said that dreams can warn us or give us joy. Dreams can have trends, especially if the conscious Ego decides to not take heed of dream signals that are important. I find it difficult to write dreams down but much easier to use a voice recorder to document a dream with my spoken word. Then afterward I can study the BTM of this audio to find the hidden messages in the dream (*).

3-3 Being startled. When there is a sudden unexplained or unbelievable event the mind is in shock. One example is to notice an animal when they are really close to it and then it darts away. Sometimes they stop to look back to you. I believe that if you listen and you have your third eye at least a little open, then you can hear the animal talk to you telepathically. This is especially noteworthy if you ask a question to it in your mind, by way of your pineal gland (*).

3-4 Hypnosis can put the mind in a state of awareness that can also access the unconscious. A hypnotist can offer suggestion that can act similar to an algorithm searching the internet. We can search the internet with a search engine and narrow a search. I think the same is true for what a hypnotist does to access the unconscious. My understanding is that the hypnotist can cause the mind to oscillate at a frequency that allows access to narrow regions in the mind not normally open to the conscious (*).

3-5 Performing BTM, as I have been discussing in this book. The subconscious is accessed via the circuitry that is in the brain. It has a large amount of connections to the ears, compared to the eyes. If you look at the circuitry to the eyes, they are basically in one bundle of optical nerves. However, the nerves to the ear are spread over a much larger area in the brain. The placement of the nerves allows more of the brain to work with sounds and therefore allows for fine discernment of sound frequencies to activate the unconscious (*).

3-6 Opening the third eye or the pineal gland. The pineal gland resembles a little pine cone attached by a stalk to the back of the third ventricle and bathed in CerebroSpinal Fluid (CSF). It can release chemicals with the CSF. I believe that the pineal gland is the center of the ajna or sixth chakra. I believe the chakras can link together to access the unconscious similar to an antenna. This antenna can detect information that is present over our whole universe. The cavity of the brain's ventricles may act similar to a cloud chamber where radiation from the universe transits through the CSF and interacts with it and sometimes causing a reaction that the pineal gland can detect. Therefore, the pineal gland could [act as] acts a detector of radiation as well as a transmitter of chemicals. The pineal gland gets hard or calcified when chemicals from outside the normal human dietary consumption are collected in the body. One way to start to decalcify is consuming olive or coconut oil in order to cleanse the dietary tract. This can dislodge toxic food products in the intestines and allow for regular bowl movements. Eating eggplant and red cabbage also helped decalcify the pineal gland, for me.[. . .]

I found in SV that: **~Eating red cabbage makes it easy to convey.~** This is in regards to the spirit communicating through the **+whirlwind+** to you. Continuing on with the paragraph:

[. . .]Eating natural products and less sugar will also help open the third eye. I found

that caffeine and chocolate help to close the
third eye and I abstain from consuming these
anymore. Originally, when I opened my third
eye I could eat a lot of red cabbage, as it
gave me purple hues when I closed my regular
eyes, but now red cabbage is hard to eat raw.
When I eat eggplant I just cut off the ends
and <u>scrub the skin in some soap and water
well then I eat the skin raw and don't
consume any of the inside pulp of the
eggplant.</u> I use soap on the eggplant to make
sure there are no herbicides on the skin by
washing it thoroughly (*).

Here, in the underline above, I found using SV the
information of: **~Matt don't balk and I hope you are
listening to this one. . .DNA. . .evolution. . .Art
var. . . DNA evolution with the RNA ah and the
pineal gland—we are all linked. . .me makes a
points.~**

Note: *Art var* is Swedish for "Species each," was
detected to be Swedish, according to my
duckduckgo.com search and a translator!

I could tell ~***Art var***~ was a foreign language
because there was a pause and then ~. . .***Art
var. . .***~ was pronounced more slowly than ~. . .**DNA
evolution with the RNA. . .**~ The speeds of
pronunciation were variable from this story-line.
This indicates that this information was pulled from
different sources. Evidently, eating eggplant skin
can activate the pineal gland and cause the RNA to
modify DNA so that we can communicate telepathically
with SBTM/BTM/SV. **~Taking vitamins is not good. . .
now those vitamins are not a food group,. . .they
make your third eye sick and make you forget.~**

> 3-7 A near death experience, I believe
> releases a large amount of pineal gland
> chemicals that result in a total out of body
> experience. The body is still connected to
> the soul by way of a so called "umbilical
> cord," meaning that you can still live
> because you are connected to the body by a
> tether. When one dies I believe the pineal
> gland dumps all of its chemicals as the soul
> permanently leaves the body. Since the soul
> is energy it must have mass according to

Einstein's energy-to-mass equivalency formula, E=MCC. A fun experiment would be to measure the mass of someone during an out of body experience to see if their mass lessens as they are astral projecting and are just connected by their umbilical cord (*).

3-8 Babies and the very young have pineal glands that should not be calcified yet. Even though they can't talk they may communicate with their gestures, the parents might be able to decipher, that the child's third eye is open and are able to access the unconscious. The American culture definitely taboos communicating to children with the third eye. The child learns this, probably through the third eye, and begins to close the third eye. The reason why is for survival because we abhor the third eye open as a culture in America, plain and simple (*).

3-9 Being with animals can open the unconscious. People who have suffered traumatic injuries find comfort in helping animals that have been trained to help people to calm their fears and trauma. I believe the animals do this because their third eye is open. Since animals can't talk with words it is the same survival instinct as the human child. For the animal however, the instinct is to do just the opposite and keep the third eye open because there is no stigma associated with other animals. Naturally this happens, since animals need to communicate these ways for survival (*).

3-10 Meeting with God or angels can access the subconscious. God told Moses, basically that, I am that I am. It was as if God was saying I know you see me here and once ago, I created the heavens, but I'm not going to give you the details of in between then and how I got here. This is subconscious or without objective words to describe the situation. The description of these situational meetings I can only imagine. Suffice it to say they will communicate without words because there are no words for their communications to you (*).

3-11 Being with family, friends, colleagues or other people that illicit memories of the past. Having others around you bring their energies to bare when you think. This changes the subject, so as to say, and the mind will change the current reference frame to the group and what that means, especially for the past. The past has memories in the brain that give support to ideas that can come forth which makes one feel that they belong. Version: 41322 (*).

Another way to access the subconscious that I have come across is metacognition. "Metacognition studies comprise another domain of research in which introspective reports are utilized in order to collect evidence about the mind (Flavell, 1979, Fleming and Dolan, 2012)." (Pauen & Haynes, p.12).

The amount of information from metacognition must be limited for each episode. Because, "In any case, metacognitive judgments only allow for limited conclusions about first-hand experience." (Pauen & Haynes, p.13). Metacognition, I think, could be done by staring. Previously on page 96, in this book, was an SV about using the two front eyes to look at someone for more than five seconds, staring.

When someone is staring, as metacognition, while also consciously looking at someone's eyes, this can conger a subconscious feeling in someone that they are looking at you with no apparent physical or R3DT connection, but it is mostly a 5D connection. The question to ask is, "What were you thinking of when you stared?"

The information one thinks can be transferred as a symbol to the 5D. The answer to the above question can also be checked by asking the one getting stared at, "What did you think that person was thinking of when they stared at you?" There is probably just one thing in their own metacognition, calling on past experience from their interactions with other people and try to gauge if they did the right thing. To try to discern other objectives of the one staring might be misguided to get the correct objective information. Because this is getting into conscious information after the fact.

To understand how this interaction between two people in the previous two paragraphs are manifested, try to imagine that looking at the things that are not conscious (in the mind's eye) can be part of a person's disposition and people just don't bother to talk about this. However, some people like to daydream for their own metacognition. For me, I think I can shine a light on this subconscious interaction between people. I started trying to access the subconscious with voice when I was in high school, as I was introverted and probably anti-social. I thought about my interaction with people and used metacognition that way. I never thought I would make SV a lifelong passion for describing the subconscious.

Chapter 4
My background in studying BTM: (*).

4-0 I first heard of playing backwards audio on the radio at about 1990 when I was listening to the radio. Soon afterwards, I modified my own tape player to play a cassette tape backwards by simply changing the rotation of the cassette motor by switching the two wires to the motor. After I graduated high school I tried to contact a former classmate, and left a message on his telephone recorder, one of my friends that might want to work with me to understand this phenomenon, he never called back. There was not a reason for me to continue as I had no reason to delve into studying this myself at this time anymore (*).

4-1 A few years later I heard a radio program describe backwards audio and the reason for its existence. Now this got me excited and I purchased one of the tape players. It came with a book of metaphors and a couple of issues of a newsletter. So I studied the method by myself and on a blog website around 1997. I remember calling a 1-800 number and asking questions and the people on the other side just laughed at me. I hung-up and thought, "What is going on.," so I continued to just study it by myself sporadically (*).

4-2 About 2016 started to develop my own thinking on how to understand the backward messages. One rule I made was to not listen to, what I now call BTM, on the weekends because I heard from the BTM not to do this. The reason is to rest, because God rested, plain and simple. Since then I have been listening to BTM as much as I can Mondays thru Fridays, excluding National Holidays and weekends. This break in listening allows me to continue quality BTM, as this break has increased my protection from overexposure to the unconscious. The main learning point to convey here is that BTM may indicate false information. The solution to this is to rest and be more discerning to find more examples of the BTM story-line that I want to prove. My rule is to do more BTM and find the same story-line elsewhere so that you can find congruence of which version of the story-line is correct (*).

4-4 My hope is that this book will legitimize BTM and allow people to start their own exploration of BTM. The main reason, or first hurdle, for not doing BTM is not taking the time to conform and adjust the hearing abilities to discern the BTM story-lines. This might be a great waste of time for some, however, in my life, I have made an attempt, not to fool anyone, or trying to write and get this information to people who want the possibility to train in this unconscious reality, probably only one percent of the population (*).

4-5 On November 9, 2021, I started to go onto social media starting with Twitter and YouTube. I was able to upload a video on a BTM introduction talk to my YouTube account that day. This is most of what I talked about: My hope with this video is to introduce backwards audio which I have named it Back Track Mechanics or BTM. The back track part is a metaphor for the train track, this is what the caboose sees during the train's journey, where the unconscious sees where the conscious goes by looking backwards to what was. What was witnessed by the unconscious on the sides and the back track form the story-line. The story-line can repeat similar to a dream and add more to it or morph into a different story. To deconstruct the story-line I think of this as being a mechanic and taking apart a mechanical device with gears and circuit boards. Story-lines are remembered by the brain with metaphor ques, so that a metaphor can call-up a past story-line and the unconscious can add to it or expand onto its understanding and details. The front of the train or engine is the conscious and it thinks it is in control, yet the conscious doesn't throw any switches to change tracks or story-lines. The conscious just adds fuel to the fire in its engine heating plant to go faster or can put on the brakes and stop. The unconscious controls other parts of the train and the symbols or scenery on the land outside the train, including the terrain on both sides of the tracks and the switches on the track. Therefore, the unconscious is really the one in charge. The conscious is but just a sheep and can adsorb thoughts as a sponge adsorbs water. Only a full or soggy sponge is not as prone to stray thoughts or water. That is one method to not being a sheep, by controlling what is contained in one's brain and keeping it full of good thought. Those thoughts in the brain have a circuit that go round and about in the brain's neural circuitry. If one can keep the circuit clean, then we can keep our thoughts and not let any outside forces or energy input their thoughts into our psyche or mind (*).

4-6 On November 12, 2021 I used the Vokoscreen, version 2.5.0, to narrate with a window on my computer desktop on how to do BTM. This made me happy because I know this makes a professional YouTube video and people will actually learn something that they can do. Wow, I should make my first video on how to install the Linux operating system so that they will be informed and use the Vokoscreen. Today is November 15, 2021, and not many people viewed my videos. I'm still thinking over the theory of BTM. I'm not sure if "I" is equal to +Me+. If not so, then it is the case that the two serving us would have to be one's +Meself+ and the +Me+. I thought that the +Me+ was a higher Self of my "I" because I thought that the +Me+ oversaw the current thoughts in my mind and could hence multitask different scenarios of the current times events. On April 4, 2022 I believe that the +Me+ is just the internal voice in one's mind and the "I" is what goes out the mouth. Version: 42122 (*).

4-6-0 When I had some money to invest I decided to use BTM to listen to company quarterly conference calls. I had a method to do this and was really careful. I made money. Then, about 2014, I got a real brokerage account and greed set in and I lost the principles I learned from before and it got really messy. I wasn't making money anymore.

About 2018, I tried futures trading with S&P 500 "ES" lots. I found myself trading the opposite to the market. When I thought to just go the other way, part of myself would not relent and I would incredibly still trade the wrong way. At one time I did trade the right way, but incredibly I somehow deep down did not feel I was worthy to make so much money.

Another failure happened. However, for those interested, I looked into using the Fibonacci levels, as those who can are probably the ones who make money. I was also looking at the oscillators and selling at the top and buying at the bottoms, when they would overshoot. I would recommend starting by funding your account with at least $5,000 to $20,000 dollars of cash and then make one trade a day.

Hopefully, trade in the morning and get out at the end of the day or the same times each day. Because once your in, you can ride it all day long if the market is going one direction. I'm writing this book to convey the basic information to others of SBTM/SV that I have found, trading "ES" futures is one way to do this. Information can come to you from an internal voice.

ES-Futures trading rules for me: (1) Only trade once you see the pattern. This might take the whole first hour to observe the market trading, (2) Do use hints from your own intuition and also from the +meself,+ (3) Use the Order Cancels Order feature, (4) Don't trade on a holiday and (5) Be capital safe.

This internal voice can include the +meself,+ ego and +ogres+ and so these may not serve me very well on their own. It is better to combine one's "ego" into the +meself+ so that they become one.

Chapter 5
The possible stigmas or taboos of exposing the subconscious or unconscious: (*).

5-0 Getting offensive responses from the revealing of the unconscious information to someone not ready for it is a real possibility. If someone can't comprehend it, based on their reality, than this would be very offensive to them. Either their unconscious would understand it in their mind, and as a result, become offended consciously from this unconscious revelation or they might get mad afterwards when the unconscious eventually reveals it to them. A result then is that they become insulted and don't know why. With this thinking they are consciously deciding that it is an insult and become offended, just from the impulse feeling from BTM. If you ask them to explain, they would immediately deny it, probably because their reaction is from associations in the brain to undesirable behavior or they simply can't understand because it is ~over their head~. This story-line means that they have not made the connection to the subconscious with the conscious. By having a reference to the subconscious then they will not take it personally but see it as a collective symbol. For example, some smells can wreak bad and be rotten so one would react. The same reaction can happen from a symbol or metaphor if it "wreaks" because this is an unconscious impulse. If one doesn't know the vocabulary of the unconscious then they will just get the impulse and can't rectify back to a normal state consciously. Imagine in our system of government if this happens, it might not work the same if subconscious information was exposed. Possible repercussions would include court action for violating someone's rights. Even though in America we have freedom of speech, that right can't be used to disparage someone else consciously (*).

5-1 In American culture we have subjects that we don't talk to other people about. This is because it is not even small talk, it is just taboo. This information is personal in nature and usually when we communicate it we will try to find common ground. BTM is taboo, so it needs rules to work with people. Those who perform BTM will get personal information and also information of value that is of a collective. Why not censoring the personal information as a rule so we have a standard of not offending consciously others in this way. This could give those who do BTM legitimacy to provide the collective information, to disseminate to those that would find this information helpful (*).

5-2 The problem with censoring is making new rules to sensor more collective information just because it doesn't suit someone's agenda. BTM information is probably used and available so now it should be collected and find a way for it to be useful. Version: 41122 (*).

I had vivid dreams one year in my townhouse when I was starting to open my third eye in 2018. I still remember these dream messages to this current day. Since I have been using my hand-held voice recorder I have learned the behavior of talking out my dreams once I wake. I have to do this otherwise I will forget the details of the dream. I can do SV on these recorded dreams and find out of there is information that my +Meself+ can tell me. So recording the dreams and doing SV on it, I think, is a good way to connect with your +meself.+ Your +meself+ can give you important truthful information, because your +meself+ wants you to fulfill your life's goals.

Chapter 6
I'm wondering if I was abducted by aliens after having some dreams with them (*).

6-0 I had multiple dreams in 2018 to 2019 when I started to open my third eye. Opening the third eye is dangerous because I made some bad decisions. However, the dreams I had seemed real, especially for the one with the grandpa faced aliens I saw. [Their] There faces looked old and wrinkled, as they must not care that they look old in their culture. I never saw or had a dream again with them as this was a one time meeting with them (*).

6-1 I call them grandpa face aliens because the skin on their faces was cracked and the skin was drooping, they needed a face lift. I was laying on a table and there were three or more around looking at me. There seemed to be a light in the room from above and it was dispersed. The light from above was just white and I remember no other details. One grandpa alien asked me if I want to do an experiment. After he said this, probably telepathically, I started just screaming as a baby. Then I woke up from the dream. The next day I clearly remember the dream. I didn't have trouble remembering this dream as it was odd and significantly different than any dream I ever had before. I told myself, if I want to do something in my life, why not just answer the grandpa aliens to their request, now. I thought they rejected me because I screamed. After I thought this, my brain indicated that I can give my answer now. I said in my mind, yes, I want to do this.

After I thought this in my mind to myself, I thought, I don't know what the experiment was about or will be. I still do not know what the experiment is. The experiment, as of now, might be partly related to or partially related to the writing of this book about BTM (*).

6-2 I remember a brief dream of an encounter with light beings. They emanated white to yellow light from their bodies. It seemed from these dreams that they were showing me that I was interesting to a variety of beings. Wanting to find out more, because these dreams are remembered as short dreams, I was also inviting aliens to visit me after a few of these dreams. One time I invited the reptilians into my dreams, because I wanted to know what they are or how they acted. Version: 41122 (*).

6-2-0 Recently in 2021, I have had dreams of human looking beings. One was with a large woman, about seven feet tall with long red hair, below the shoulder. She was interested in my bathroom, and asked, "Why don't you show me your bathroom?" It happened that the next day I was moving out of my apartment and I left my bath towel hanging on the shower curtain rod. If I did not go back into my apartment and the bathroom, as I cleaned for moving out, I would have never got this towel back again.

I have talked to a hypnotist, to rent an office next to hers and she wanted to know if I had any thought as to reasons why "reverse speech" was so, or other conflicting opinions. I sent her an email detailing a duckduckgo.com search on skeptics of "reverse speech" and then my ideas against them. She just commented, I believe, that her insurance could not involve me with her open room as a rental.

Chapter 7
Skeptical Arguments (*).

7-0 Some will say that the spoken words heard from anyone when talked backwards are just there most of the time because there is an artifact to the speech from the way it is pronounced forwards and just happens to make a word backward. But the artifact has no purpose and is just there by chance. On the other hand, some words are the same forwards as backwards as in "race car." Surly these words or palindromes will be heard backwards because they are spelled backwards the same as forwards. Some words sound the same way when pronounced correctly forwards and then so sound the same way backwards, each time. This is because the word is pronounced exactly phonetically forwards, of course the backwards will be consistent each time also (*).

7-1 So what, "How is this going to improve who I am?" Then they ask, "Can this be proved from something that is not scientifically proven?" They might think that it is a waste of time and money. I would say, "Look at your children and listen to what their observations, comments and questions are when they come into this world." The reaction of children is unbiased because they are constantly learning and have no need for hiding the truth or obnoxious opinions. They may say that, "This challenges my world view and I want nothing to do with it." Version: 41222 (*).

7-1-0 However, as some words sound the same each time, there are times when we are in a hurry or something happens and we don't pronounce the words the same each time. So that, there is the impetus from the brain that can modify the pronunciation to suit the SBTM/SV information needed by the unconscious to get something out of the mind and into the CV. I imagine SBTM/SV as a frequency modulation of the regular CV, than the usual sounds for a word that sounds perfect in pronunciation, the same each time.

I believe speech can be unconsciously modified backwards to add information at the same time with the normally speaking tongue, because this is how the universe is made. Someone may ask, "Is this true?" I would ask them, "What is truth?" I would respond that logic is sufficient to argue for the existence of SBTM/SV because truth is just the sides of the coin, <u>there is the space between the coin that determines what side the coin lands on. The truth is</u> . . .

Here, in the underline above, I found using SV the information of: **~Such is the illusion . . .wasneack. . .where the coin stops is quantum mechanics is safe, that's weird.~**

Note: I'm not sure what ~wasneack~ is, but it might be about the middle of the coin adding to the choice of side-landings is clearly wrong and is an illusion that makes sense. So a **+quantum+** is **+safe+** to say, that it decides the side the coin stops on, continuing on with the paragraph:

. . . just the end result and the nature of SBTM/SV is that it has a hard time being specific, so therefore the truth will be sprinkles in SBTM/SV, however SBTM/SV is **+all-encompassing.+**

The meat of SBTM/SV comes with the realization of the possibility and where this possibility leads one on, is a journey. Some may not appreciate this journey, some reasons may be that: their ears are damaged and cannot hear the SBTM/SV and just hear minutia or their belief system cannot allow for these beliefs that give the belief for SBTM/SV to function with their third eye, even if the third eye is just a little open. Therefore, their chakras are not aligned and they don't have good intuition and then they might gravitate to apophenia.

Some people have listened to **+reverse speech+** and tried to memorize each sound and then when they record it and reverse it, then they sound as forward speech or CV. I would say to this that they are using speech as a particle and objective where SBTM/SV is the wave and has a meaningful response to one's awareness to unconscious thought, not just single words, this is subjective. Because SBTM/SV is subjective, it takes an interpretation, because since the mind communicates in metaphors (figure of speech) and story-lines there needs to be a translation from the subjective to the objective, which can be hard and time consuming. The interpretation might just be in me and not a phenomena of our Universe. SBTM/SV might just be in the mind of the individual, as many may have suggested.

Besides someone telling me I'm wrong, it's just in me, they may say, to relent and face the music. They conjecture on the mode of transportation and the use of the subconscious to transmit information in speech, that can only be in one's mind.
According to some, the subconscious is not testable only opinions. Just look at how people can be programmed in their mind unconsciously in the media and from outside the mind, it's not just individual.

Chapter 8
Unconscious programming to one's mind (*).

8-0 Unconscious programming can come from cues in the environment. I consider unconscious programming as possible hypnotic luring to take advantage of another person. This can include what is said to you, in comments from another person, the news media's take on a news event, unpropitious occurrences that happen around you or in your observation and things you tell yourself consciously. We are after all, as if we were sheep, easily herded away to some musing. What these unconscious programming can do is to change the conservation topic maliciously so that you begin a new conservation not in your best interest. As a result, you are the conduit to spread the results of the programming. This can be repeating what someone else told you, ask yourself, "Is this true?" You can under these programming give your own personal information that you want to keep private from others, ask yourself, "Why does this person need these details?," quickly thinking I thought, "While, I will just tell general information and not specific details." These events can happen and the foolish will not even know they were fooled (*).

8-1 Unconscious programming, from the outside of the selfs is just an environmental reaction to the person, in question. The person observing an event may not be examining what they are looking at to figure out how this will effect them. An option one has is to remove themselves from the situation and then to determine what is the root cause of getting programmed with this information. For example, if you have friends that want to bring you into a bad situation because there are risks to your future in society then the easiest choice to make is to get new friends. Some outside unconscious programming include ethnicity and culture, for instance in America during the October Halloween festivities people want to go into the haunted house to get scared, this is repeated as [a] generational unconscious program. The place one is at can be programming your actions without you even knowing it, for example in a church when everyone is praying they bow their heads to be reverent to the prayer in the holy place. This place then is an important aspect similar to ethnicity and culture in the power of the unconscious, that can make us do things (*).

8-2 A place may have energy associated to it in the ground or terrain as in topographically which can include mountains, hills, valleys or energy associated in the design of a building or structure which can include the people working inside it. A person may affect me as a psychic vampire or how information is marketed to my eyes and ears as in viewing television commercials. When the mind's senses are available then they are on. Conscious means that the sense will be consciously delivered to the mind on demand, it can't be turned off, the thought can only be remembered in its truth or diverted to another memory to subdue the offending sense if it is false. These occurrences are censored by my beliefs, conscious thoughts as well as my unconscious framework and the society I live in. I believe the unconscious is built: on past lives, because I found my first previous life with BTM; on the reality of one's situation in life, including their place or class among society and how they commit to change consciously. Because the current life of the soul needs to be recorded or connected in some manner into the unconscious for it to be transferred to the next life. What I see and hear is hard to turn off once it is received and understood by me consciously. Disciplined use of our senses will allow us to remain calm under stress which is important if we want to find the truth. No wonder, when practicing yoga that we want to center the body and soul away from our immediate surroundings to access vital information of the universe, the collective unconscious (*).

8-3 Sometimes unconscious programming sets into one's own psyche and whatever situation they find themselves in they still make the same mistakes. That is how BTM can help, it is to reveal to someone of their unconscious programming so they can ask the question, "How can I avoid this happening again?," because initially this person may feel that it is karma and they are destined to continue on this path. I say since we can choose and make choices and others that see us will see the choice and say, "Well, you choose to do that!," rather so, that we can also make a choice to change as a person so that this karma no longer exists in ourselves. Our routines sometimes hurt us when we are too friendly to people in our neighborhoods and a stranger comes around and takes advantage to try to get a survey on an important issue, for example. All they want is your information so they can get paid for it somehow. BTM should be used so that what details are presented in BTM can be shown or give a practical solution. Write a script on how you intend to change and then read it and then do BTM on it so you can see any patterns. The patterns might be dead ends that just lead to the +caldera+ or think positive and examine other issues to see a broader pattern in one's behavior. If there is a dead end or +caldera+ then consider revising your script and try BTM on it again.

"What do you do if there is an evil presence around you?" Ask God for help and keep doing the right things. Evil is in this R3DT reality just as good is. We are going to find evil. I suggest not to look at a blasphemy and ask God to take these corrupt thoughts away from you. When some evil is acting on you and you can figure it out then examine what the trend is [and] ask if there is a solution or common sense action that you can do to avoid the injury or sin from this. In the future, things should get better if not then, consider making a script from the Bible or other holy book, as in the Lord[']s Prayer from the Bible, or just ask God for help [Mat. VI:9 ABS, 1858 &] (*).

8-4 One way to stop an unconscious karma programming is to consciously be aware of it and give it a designation. As an example, to tell oneself in a situation, once the unconscious event makes itself aware consciously is to then say to oneself "I'm going to think about something else to divert the attention away from this hostile point." Sometimes, I can imagine that, one cannot immediately remove focus, as in the act of driving a car, but must eventually, when safely parked, so one can know the programming and divert attention away while there are no distractions. Divert attention to the patterns that led to the situation and know those patterns and what those indicators can lead to. There should be a sequence of events that you can flow chart to get to the event, that is the unconscious programming source. Part of this flow chart choices are to go to a different place, get new friends, move your focus to somewhere else if you can't go to another place, don't take part in conversations that are low in energy. Only respond with facts, if you don't know something then just say, "I don't have an official answer for you." Sometimes others with low energy will try to bring others around them to ease their pain and hurt so they can feel better and then look at seeing someone else suffer too (*).

8-5 If I want to change my life for the better then I need to realize I need to shield myself or divert energies that come toward me so I'm not destroying my chances to succeed in these goals. This might result in loosing friendships, however, I might find some new friend if I live the life that I want to try this lifetime. How to do this is control what you see and hear. Since when I might see something horrible, I might put my hands over my eyes to shield me of the pain while looking at it. The same for my ears, I can put my index fingers into my ear canal holes if there is an offending very loud noise. How to divert my focus when this happens is to, think of the past, future and present. Try to recall if this happened before and what worked best to solve the problem. Then see or imagine where I want to be in the future through my goals and rules for life. I'll ask myself, "Do I want to break a rule for a small [thing] ~~think~~ and risk loosing the big thing because some outside force is tempting me?" Now, I'll adjust to the present. A good option is to close my "clam shell," because talking can release subconscious BTM information in my speech and other's unconscious will pick it up and may react to me. When you are isolated, this is the best time to start over and make rules to avoid past mistakes (*).

8-6 Don't react, walk away or refrain from conversations that are not productive in the moment. A good example is the +quicksand+ BTM. Given a problem there are correct ways to solve the problem and wrong ways that will sink me into the quicksand further. I believe if +quicksand+ is heard then don't make any commitments or big decisions. Try to put off what someone suggest is an urgent issue, this may not be true. The issue could be put off with some understanding of what is affecting me, either physically or mentally. Time can cure the +quicksand+ metaphor.   Version: 41322 (*).

How do we get programmed in the unconscious might be related to how we think in general. The thinking process might begin with a symbol that the mind's eye can see. This image becomes an enticing suggestion to direct a story-line to happen. Having a logical thought requires us as humans to think of the consequences of actions and the behavior prompting it. Stories from wise individuals can teach values and principles. These stories can be remembered so that the student thinks of the story and internalize it. Once internalization of a story occurs, then people think that they thought up the story. They can change their life because they got smarter. Stories can change behaviors with the subconscious.

Chapter 9
The nature of cognition, both conscious and unconscious (*).

9-0 We can think of the universe as a way to promote thought. Because I believe the universe is thinking, at least, in the realm of the unconscious. At the very basis of life is thought, that is why I think it is dominant across and also outside the universe. This being so, because outside the universe there are not even dimensions, I think. The dimension is a construct of those who created the universe, God. Outside our [U]universe, I believe, there are elements instead of dimensions. The elements are, or could be, I believe, are related to thought. There are different flavors of thought, including conscious, unconscious, logical possibilities and illogical possibilities. I describe the third type, Logical Possibilities or LP, as thought without anyone there. LP involves the nature of form and function of being. Just being there requires that there be an existence to this form. The form has beingness because as we all know there is a thud in the forest when a tree falls down and no one has to be there to hear it. LP are just there, part of the fabric, because the physics of this universe is constant. I believe outside the universe this may change. Because if thought can be dominant then illogical things can happen, that is evil, at least from our reference frame in this universe (*).

9-1 An extraterrestrial, from descriptions of abductees, can go or are reported to go through walls. This seems to defy the logic of physics as we humans know it. "Could there be a realm outside our three dimensional space?" I think so, because if our real three dimensional reality is moving or oscillating just beyond the boundary of the real three dimensions then there might be something outside it as in dark matter that is causing the oscillations. Dark matter has dark dimensions to hold the dark matter. Having dark dimensions means that there is simply no light. Doing [an] a~~ ~~ OSA here: if we have dark dimensions with no light then this is similar to a black hole. A black hole also has no light. A black hole has no light because basically one of its ~~the~~ dimensions is contracted to a point that the black hole is a disk, as seen from in the real three dimensional space. It is as if one of the dimensions of the black hole is curled in onto itself as a cardioid or circulate with angular rotation that disallows light out because light travels in straight lines. The black hole would then be the kink in the cardioid. So if dimensions curl then there is no light. Therefore, in the real three dimensions there is no curl present, we have to design a device to make curl. <u>We can create curl with collecting</u>[. . .] (*).

Here, in the underline above, I found using SV the information of: **+me under warfair Matthew,+** I think I might me entering a downward spiral, again, and ETs or space aliens might be conducting psychic warfare on me because I'm exposing them. Continuing on with the paragraph:

[. . .]mass or energy to a point. We can focus energy to a point if we create a device similar to a Tesla coil in the shape or design of a cardioid. Usually, a Tesla coil will make a magnetic flux shape and not bend around to make a cardioid, but what if the tube for the Tesla coil is cone shaped, as if two cone tips are ~~places~~ near each tip, a cardioid shape might occur in the flux lines. Just look at the magnetic flux lines of two opposing magnets, they seem to make something of the cardioid kink when forced next to each other. The frequency to use would naturally be a harmonic of the frequency of the universe. I believe the size of the universe is static and so the frequency would be constant which would lead to a very slow frequency and I suppose the harmonics of that frequency would be many orders of magnitude greater. Unless there is a device with a very slow frequency output, but I'm not aware of any such device besides a Tesla coil (*).

9-2 There may be a way to transcend our three dimensional reality just by thinking the right thoughts. Because unless we have technology to curl space in our real three dimensions we may have to resort to our mind to somehow do this in a different way. Similar to making the pyramids with brute force and leverage instead of by levitation technology that uses antigravity. If outside the universe there are elements of existence that are measured on thought then the thoughts we think can somehow associate with outside the universe with those elements. The result is a flow of information to oneself that would be spiritual in nature and require chakras that are in tune with each other and the higher frequency of thought than the average [conversation] ~~conservation~~ on planet Earth these days. Supposedly the energy of the chakras can be centered by breathing in deeply then exhale quickly. You should feel a tingling on your skin all over your body. Once in this state then one can ask the question, "What greatness of higher frequency is outside the universe and how can I obtain it?" One will probably just hear, "Read the bible and understand its meanings" [ABS, 1858 &] (*).

9-3 There may be a way to think of thought as pervading the universe. Thought in the real three dimensions cannot be seen. It is as if thought here is as a gas, or see-through with no substance. Thought in the higher level dimensions or the 5D could be imagined as a liquid. Because God is in the higher dimensions and can see good or evil from far away, as thought could be thought of as a river. One can see a body of water in 5D and avoid it if it is bad or go and join it if it is good. In this way God can, I think, see from the distance the intention[,] where as we in the real three dimensions we have trouble seeing bad thoughts or even have trouble identifying when we are awash with bad thoughts till it is too late. Version: 41322 (*).

If one can tap into the subconscious then let's make a way to program the unconscious cues and program them to what we want, including success and prestige.

Chapter 10
Motivational tapes for one's mind (*).

10-0 Motivational tapes can be made that put one's own subconscious into a background of audio so as to sooth oneself into relaxing and allowing oneself to open up to the possibilities that the subconscious can help them. I created one with an effect of my audio sound editor, with a sound that reminds me of the ocean waves hitting the coast and my own script about how to relax and have supportive energies spoken to revive and use one's day to the fullest. One script I thought up was relaxing one's body[,] another for stop smoking, the ideas here can be focused to the client's need for what they want to stop or control. The idea is to bring to light a mentality of mindfulness for remembering the consequence that can occur and act to cause oneself to change for the better (*).

10-1 Motivational tape script to relax one's body: Please sit down in a comfortable position; imagine you are on your yoga mat and you want to relax and want revitalization of yourself; you see that you are going to get relaxed right now and you can relax all parts of your body; relax your neck, you can turn your neck to a relaxed spinal position each way; relax your arms, as they are filling with Earth energy; relax your upper torso by taking a deep breath, and you are exhaling and feeling more revitalized with Earth energy; with each deep breath your chakras are slowly synchronizing to the tune of your breath; relax your lower torso by turning to the left and turning back to the right; relax your legs as they are closest to the Earth energy and wiggle your toes as they are the easiest to relax; light codes are flowing into your crown chakra and down your spine to its bottom and into the Earth and then out to the universe again; if you could see yourself now you are an electric and magnetic field connected to the Earth's fields of energy as a torus; you are one with Earth energy and all tensions are getting grounded to Earth leaving you refreshed parts of the body one by one; you feel this happening and can visualized the parts that are easily refreshed and those that need more help; let the parts that are totally relaxed give energy to those that are not relaxed, fully; relaxing now to help the chakras synchronize so that your intuition is connected and strong; each chakra is linked together as if by magnetic fields and connected pole to pole; finally you have the support of the Earth and it totally supports you; now you will start to get up and stand straight and take a deep breath and feel all parts of the body as one and exhale to release the Earth energies that filled you up and with any remaining tensions out of the body and into the universe (*).

10-2 Motivational tape script to communicate to friendly starfriends: I want to be contacted by friendly starpeople; I want to know if I'm a starseed; I want to be beamed up to the mothership tonight and remember or at least ask questions telepathically about the experience when I wake up; I know there are barriers in my society that allow disclosure to be impractical, but I want a personal experience or a group experience tonight with like minded Earthling people, such as me, to explore the future of mankind and know the problems ahead; I realize that the starpeople may be from my own future here on Earth; this realization is curious, because, "Why change the past if there are logical problems of changing the past?"; perhaps the past is an illusion and we are all on a set of dimensions for each history as there are ever changing possibilities as forks in the road to the future, which might look as a big haystack; that is why I want to communicate with starpeople that are friendly to the human race; I want to see that haystack of history and understand that the future may have the same ending for all of us or if this is not true then to let me know; I want to know what will happen when starpeople are recognized as real and what will happen when we wrestle with new technology and I want to know how we should prepare our body and mind to be with the starpeople, including how we should think and interact (*).

10-3 Motivational tape script to stop smoking: I'm imagining my lungs opening up as two sides of a deflated rubber inner tube separating as they were stuck together; I enjoy filling my lungs to capacity with outside air, crisp cool mountain air; I'm quitting smoking and will replace this activity by using my smart device, doing a workout or an enjoyable walk, starting out slow at first and then build to the level I need to live comfortably without smoking; there will be lots of choices to choose from to replace the ques and symbols that made me smoke; I can consciously identify these and replace them with healthy activities instead of smoking; I have a plan to stop smoking and will keep an open mind that if I fail then I will continue to prevail by stopping again; eventually, the harmful acts and thoughts I have that made my life unhealthy will wither as weeds in the hot sun with no rain, weeds in this situation will die as will my habit to smoke; all I need to do is take one step at a time and one day after another and my goal of healthy living will flourish because I have so many options and ideas for living healthy and they are close to use as is the tip of my tongue or bookmarks in my internet browser. Version: 41122 (*).

10-3-0 Motivational tape script to make money trading S&P 500 future lots, symbol ES: I understand the players that trade ES and understand their leverages, ways that, and reasons that they trade; I'm a smaller trader and rely on intuition and SV to understand when to trade and take profit; I can listen to my higher-self and obey his/her direction as I know if I trade with wrong intuition, then I will loose all of my money; the only way to survive trading ES futures is to follow the rules; these trading rules include using Fibonacci numbers for end points to know when the machine algorithms will trade; I will change the way I traded from before and place trades that can ride a long time as I know this is the safest to do with the volatility of ES price discovery movements.

I will trade a buy always when money is going into the market and sell when money is going out of the market; when I hear that the **+meself+** hears **+morphine,+** that is when I will stop trading for the day; I will stop, because to trade at this point is to loose money; my mind at this moment would not be capable of receiving the impulses of unconscious information; it is as if the faucets of the chakras are closed and the information is not forth coming and dry. It is dry, as described by the Bible parable, as the seeds that, "[. . .]they withered away." and be eaten by the ravens and other "[. . .]fowls came[. . .]" because they are just thrown and scattered by the wind; I can trade with paper and not loose money; therefore, I will not trade if I'm not in the mood of trading as there may be outside forces giving me a psychic attack, so

that I will loose my money; I want to make money, I
want to make six figures trading as I know that this
is possible to do; there may be reasons why I loose
money as in my past behaviors and they are from my
choices I made; however, now I have experience
trading ES and I believe that if I give myself a new
chance, know my **+Selfs,+** can conceive of it, then I
can make the right trades and make money
consistently (Mat. XIII:8 ABS, 1858).

10-3-1 Garden of Eden: Thanks to the Bible, there
are stories from the ancient days, one example is
the story of "[. . .]man[. . .]" (Gen. II:8 ABS,
1858). I believe that the first humans lived in a
different set of dimensions than the R3DT that we
all live in, before they fell. Why? Because there
was a fall from grace. I believe that Adam and Eve
were removed from a higher set of dimensions and
were placed into a lower vibrational state because
there was sin and this resulted in fear for one's
survival in this space. The cherubim do watch and
are protectors for access back into the Garden of
Eden.

I believe that the entrance to the Garden of Eden is
through a worm hole into another set of three
dimensions. It could be timeless or another form of
time residing in there. Perhaps time is not linear,
but nonlinear in this space of higher frequency. If
time is not linear then one might not need clothes
because it would never be too hot or cold because
there is not linear time to dissipate heat.

Furthermore, this space would have a quality of a
dream where reality could morph into another
reality, just as one story will change into another
or a rotation into a **+whirlwind.+** If this happens,
then there might be other dimensions connected to
the higher three, that Adam and Eve were in, that
allows for this abrupt change. I think the worm
hole is connected by zero points to and from the
R3DT and the 5D.

10-3-2 5D Transformation: Imagining this higher
three dimensions or 5D and all of a sudden something
blinks into the field of view. If there were five
dimensions as this higher dimensional reality, as in

5D, then we have a higher three dimensions connected with a 2ID worm hole. This worm hole, a two dimensional worm hole, is what allows for the sudden transition into a different reality. One may ask, "But how to get to 5D?"

The answer is through first controlling one's thoughts to be in a higher vibrational state. However, since the flesh is sin then only the mind can go to the Garden of Eden or 5D. One would have to have a different body, assuming they had the mature mind, to enter. 5D might be comparable to the Garden of Eden.

10-3-3 Snake people: There might of been animals in the Garden of Eden. Snakes and birds all have been represented as symbols in Egypt's tombs. I think alien types are in modern culture, especially those that have dreams of these or were abducted by space aliens. The snake people were caused to slither on the ground from what happened to them in the Garden of Eden, because God's commandment was broken. Adam and Eve were able to communicate with the snake and also God. Conceivably, the other animals in the Garden of Eden also could talk since they are in the 5D.

10-3-4 What Adam and Eve did in the garden: I think Adam and Eve were helping God tend to the garden and all of the plants and things that grew there. They were told not to eat of the tree of wisdom and life. Adam and Eve were without sin, meaning that they had no knowledge of breaking any commandment, their thoughts strayed away from even contemplating breaking a law, for they were focused on helping God tend to the garden and would have a long life.

When God rested, then they rested. I think that Adam and Eve were then in constant connection and knew God intimately. When they ate the apple from the tree of wisdom and life they lost their life, in the Garden of Eden, because Adam and Eve understood a different frame of reference, they gained wisdom.

This frame of reference caused them to see themselves in a different way. This new way made them hide from God because they wanted to do their own thing away from God and therefore in this 5D reality, they were as if they were gods. Since they were in 5D their thoughts became reality because that worm hole was opened. It is as if the worm hole was curled open in the 5D reality, but in our real 3D, the worm hole is curled closed, since we can't see the opening to the Garden of Eden.

10-3-5 Assuming there is a set of five dimensions, the 5D: Adam and Eve were able to live in the 5D with its discontinuous, nonlinear time or rotational frequency. Time in the 5D could jump to any time that has a presence in the R3DT. Hence, time in 5D would be the same time present in the R3DT, but the worm hole acts as a time machine.

Thought can manipulate the worm hole, since it is always open in the 5D, and transport the individual to any time. However, since the Garden of Eden has boundaries it is smaller than the R3DT, but never-the-less has the lens of the worm hole to go anywhere in the R3DT. Therefore, size is relative and that may be why the "[. . .]giants in the earth[. . .]" were so tall, without explanation (Gen. VI:4 ABS, 1858). Now if there is a 5D and the R3DT, then there are three dimensions left to total the eleven dimensions of String Theory, because five plus three is eight.

The missing three dimensions could encapsulate the other eight so as to separate the universe from whatever is outside the universe.

Chapter 11
Using BTM for self-help (*).

11-0 I remember being in the [principal's] ~~principle~~ office when I was in grade school. I was there because I was not giving in to the teacher's instructions. I was yelling and screaming in the classroom and including while in the [principal's] ~~principals~~ office. The principal just left me alone and I was left in the office and essentially screaming at myself. Eventually, I got quieter and quieter until I was shut-up. "What caused myself to lower the volume?" It might be the thinking in my mind that, "I can make no claim or prove to anyone that I'm right." I had nothing to argue at. While there I was talking subconsciously, I think, that caused me to intelligently consider other options to solve my problem in the principal's office. "What if I could of heard my BTM back then" (*)?

11-1 Listening to one's story in audio can document one's problems. Better yet, do BTM on it, so that one can hear the subconscious information. Just listen to the whole thing, one might fall asleep after connecting to the brain waves. After one realizes the problem they have from hearing the story-line in the BTM there is a sort of euphoria of a connection taking place that allows energy to be neutralized and making the outlook of this person better. The only problem is, "Are they going to get in this same problem again" (*)?

11-2 If they find themselves in the same predicament again then they did not change. Change requires access to a memory of the past results. This is similar to how we try to access long-term memories. Sometimes we can access these memories, other times we can't. For long-term memories we can't remember, we may find ourselves just sitting there asking questions, "Why don't I remember this?" A solution is to find something about the block in the memory, to try to remember it [in] a different way, making a circuit around and down different neurons in the brain to bypass the block (*).

11-3 I believe listening to BTM is a tool to find alternative neural pathways for understanding a problem. Because in the process of doing BTM one needs to open the mind to the possibility, making a bigger net, so that stories will form from what those sounds can be identified as, for the one listening to it. It can confirm what one is already thinking or adding to new information that might be hard to understand. The

information that is hard to understand will
just hang there in your thoughts unable to be
dealt with or be useful, until there is an
impetus to put things together and realize
something new. Version: 41322 (*).

11-3-0 I believe that SV can allow one to make an
audio track for self-hypnosis. There needs to be
something to study and try to find the answer to the
question that describes something to be solved for
them. By finding a problem and analyzing the
different solutions to that problem, in one's own
words, then one can find the +S-M+'s that are
programmed in them and they change this programming
with SV. When one stutters this might be the
+meself+ trying to speak.

11-3-1 I was working at my job and kind of tired
because I just needed to clock-out. Since I was
starting to fall asleep I was not applying a lot of
brain power to my surroundings. Another person
walks by me and says, "Hi." I didn't know what
exactly to respond with since I was not paying
attention, but I stuttered out something about how
it was going. The person quietly chuckled and then
I tried to repair the situation and said, "Have a
great day," to the person.

I believe since I was basically not thinking that my
+meself+ responded in the garbling that came out of
my mouth. Since I was startled, I did not
consciously say this soon enough and tried to repair
the conservation by saying something nice back to
the person. My thinking from this experience is
that people who stutter are giving their **+meself+**
unconditional access to their voice box.

To test this hypothesis I would have to do SV on
people who stutter to find any clues. Nevertheless,
if someone who stutters would do SV on themselves
then this might open the door for the conscious mind
to detect the **+meself+** and then integrate the
+meself+ back into the area it is suppose to be in
and not through the CV component of speech.

I should warn those who may have a reaction to
finding the subconscious of oneself or others to the

possible consequences. The reaction might consist of a head-ache and irritability from knowing underlying reality of the subconscious and unconscious if they are not ready to deal with it.

Picture 2:

Back Woods Roads to your Past Life

Chapter 12
My Disclaimer for those interested (*).

12-0 I won't have many sources in this book, because if I do BTM on someone I could get sued by them. They would say, "I did not say that!" and sue me for disparaging their reputation. That is the reason BTM is taboo, in my opinion, or people would fear it. I'm just trying to get the word out about this information reservoir and it is available to us as humanity. I hope humanity can take advantage of this information, less artificial intelligence and machine learning takes the advantage first, to better all people in understanding each other. The two people I do acknowledge I believe need to take the credit for what they did. I honor them in the Acknowledgments in this book and the two acknowledgments are: (1) I did not create the term unconscious with the Id, Ego or Superego, that was Sigmund Freud [and] (2) I did not create the term Collective Unconscious, that was Karl Gustav Jung (*).

12-1 Use BTM at your own risk, ~doing **** [+reverse speech+] is hard, it makes most people insane~. If they take a break, one or two days a week then they might feel better in the long run. Your analysis may not always be right. My analysis may not be the fact. The state of the mind can change between listening to BTM and therefore change the story-line. Please consider anything punctuated by the plus [for] ~~or~~ metaphor (+), tilde [for] ~~or~~ story-line (~), or carrot for [subconscious] channeling (^) as entertainment in this book. My recommendation is, try to find out when BTM is correct and do BTM in that way. I do not take credit for BTM. [I do take credit for SV, which I think is real.] I do not take credit for using the thinking method of OSA to specifically decipher BTM, as I have done no research in this for what others may of done. Through my BTM method, I claim, that there is an interaction of the cerebral cortex with the cerebellum. With the link to the cerebral cortex with the cerebellum there is a way to perceive a transform of the unconscious dialog of selfs into a consciousness realization. This is basically decoding what was inputted into speech spoken from the tongue. I do not take credit for this because this was from the unconscious and therefore anyone else could find it too (*).

12-2 Opening one's third eye or pineal gland can change yourself. This change can set different goals from before and you may now

want to have rapid change. This can lead to
making bad decisions or other disasters if
the consequences are not thought out or
written down and examined. Even though you
may feel you have good intuition you may not
be able to wait to see the whole picture,
especially if [your] ~~you~~ combining opening
the pineal gland for the first time and also
doing BTM, as the chakras might not be
aligned and you will have bad intuition (*).

12-3 This book is not medical advice. The
reader should seek professional help and not
rely on my theory or advice to help them with
any particular health or mental issues.
Version: 41322 (*).

12-3-0 I have certain values as a professional SV.
Sunday is the sabbath and I try to keep it by not
purchasing or doing SV. Just visiting and
reflecting is good enough. I choose not to use a
particular word, but will refer to it as a reference
that is in the Bible. "[. . .]any man hath a ||
running issue out of his flesh, *because of* his issue
he *is* unclean." (Lev. XV:2 ABS, 1858). I will use
this Bible verse as equal to the term-X.

Another profound discovery for me is that doing
"term-X" can mess-up the chakras. I have even been
informed from listening to my SV that the devil can
take energy from me when I do "term-X" and this
might be a cause of downward spirals that I've had.
Moreover, that when "term-X" is done, it is as if
there is a plug removed from the bottom of the root
chakra that allows energy to enter the chakras, that
includes energy from Satin.

Even my thoughts, I've been made aware, can have
significant effects when making every-day choices.
The choice can be skewed by the thoughts of the
+selfs,+ even though they are consciously forgotten,
they can remain stuck in the subconscious ready to
act without conscious control! This is why, if one
has an offending image or speech enter into their
mind then this trigger can be contradicted by
thinking of something nice, as in nature.

This nice thing can be specific to the individual. As an example, I think of a running river flowing down the creek and then over my body that washes away the offending idea. Something artificial might not be able to do this relief. Can I say that, "Thinking nice can change one's thoughts."

12-3-1 <u>"Can you see your thoughts?"</u> **~Matt, sorry you can see your dreams right here you are false.~**
You can see things in your neighborhood and indicators that manifest by using street smarts to predict the outcome. However, since we are asleep we cannot see thoughts (perhaps we can see dreams and these are from symbols, from the above, or hear the thought with SV) and at best see symbols in our dreams with our third eye. We may be able to organize our thoughts to have a better memory.

To know what our thoughts are, we need a baseline thought from our brains, when we are calm and alone, to calibrate to. When we get irritated then there may be evil thoughts that came into our head by way of the **+whirlwind.+** The spirit comes into us and causes us to pass through some **+wind+** by some force that is leading someone into some predicaments, as a **+snakehead+** can do. It is said that the **+snakehead+** is the smuggler that brings a migrant into a promised country. This can happen because we follow our thoughts even though they may not be ours. The baseline can help us discern the feelings from a thought and therefore keep us on a narrow path, and not deviate.

12-3-2 Media and marketing that bathes our senses can be subverted by the devil, all because it is a lower energy level. I feel lower energy when attacked by someone else. Some of my civilization leaves me from my conscious senses and I'm reverted to fight or flight. Fight or flight is not wisdom, it is just evolution. Matt asked, "Will I make the right choice to confront someone that attacks or just leave? This is the open question."

If I fight then I go as low as they are and if I leave or ignore, then I traveled to a different level. I want to be at a higher different level, sitting on my own pod of wisdom. However, this will divert me from living in the world. Why not act to help friends or foes while also acting at a higher frequency, sort of similar to how a superhero acts. The superhero doesn't react to bad lines or jokes in a movie, he or she just directs their energy to doing good for the whole of humanity.

12-3-3 I recently had this information received by me that there is an organization run by humans, and they rule the money system. I have found that they might be the ones conducting a psychic warfare on me, so that I do not make money on my investments. The reason why this is, might be, is because they can distort my thinking.

Distorted thinking can come about by damaging my chakras and this can allow for bad intuition. Well, I'm starting to see through this, from trial and error. The best way to fight it is to eat right, keep a daily routine, don't break the sabbath and know the consequences of one's thoughts so that they are not dragged down into the lower energies of the R3DT frequencies.

When I listen to SV, I can become sleepy. This is especially true when I'm already tired. When this happens I need to become consciously aware of the sleepiness or else if my head is up then I might make a whiplash as I nod off to sleep. To remedy this I just lay my head on the back of my computer chair so I don't cause this trauma to my brain. I'm convinced SV is a true method to access the subconscious, as this is really happening to me and useful, and SV is not for entertainment purposes because I did publish this book to get SV information out to the public as a real concept.

Lastly from my old website, I gave respect to where it is due from the efforts from others, I should of also included the founder of Reverse Speech Technologies and his followers, I don't do this method and I don't follow these rules, so I do SV.

Chapter 13
Acknowledgments: (*).

13-0 Thanks to the USA, there are probably countries in which I could not publish this book in. Also to Sigmund Freud for his work showing the unconscious forms of the Id, Ego[,] and Superego and then Karl Gustav Jung for his work in the Collective Unconscious. Version: 41122 (*).

Chapter 14: Introduction SBTM/SV experiments

14-0 Conducting experiments to help understand connections from one's reality to their subconscious. Follow common accepted testing protocols, for example, by asking permission to test someone's subconscious.

1. Do a generic experiment in different ways for relaxing and then testing SV. Examples of relaxing prep include: yoga for 15 minutes, as in self-relaxing every part of the body with feet crossed; consuming a set amount of red cabbage and eggplant for 24 hours before being tested, after waking-up the next day; self-hypnosis with a pendulum for 15 minutes; listening to favorite music for 15 minutes; listening to brain activation frequencies for 15 minutes; or sit in a dark closet for 15 minutes trying to imagine opening the pineal gland or third eye. After these relaxation methods, the third eye should be relaxed, since the body is relaxed, then conduct an experiment by listening to a recording using the SBTM/SV method and testing and comparing brain waves as well as SBTM/SV comprehension.

Use brain sensors to record brain signals from the skin as the subject is listening to SBTM/SV, for finding SBTM/SV aptitude indicators. Allow for a quiet uninterpreted room for this one testing to listening to a standardized recording for use in SBTM/SV research.

In a standardized track, knowing the times when the SBTM/SV messages occur, see if there are any correlation to the brain waves at the same time intervals as metaphors or story-lines for multiple subjects, that have never listened to this audio track before. After multiple subjects have performed this experiment, then test to see if there are any general brain waves that prime the mind to pick-up on SBTM/SV. Here we would be looking for an easier way to listen for SBTM/SV.

2. Doing a recording of someone's casual conservation and then do SBTM/SV on it to find the hidden personal information about a person. Then ask that person questions about its validity.

3. Have someone write down their best days and worst days of their life. Then have them record while reading these stories out-loud to themselves. Afterwards, listen to the SBTM/SV and see if there are any congruent information. If other information comes out, then ask the participant to help explain it, as done in Experiment 2.

4. While someone is dreaming, in low-frequency REM sleep, wake them up to record their dreams. Quickly have them do the recording for an SV and see if there brain waves are the same with the one's primed for SV from Experiment 1.

5. Try to find the Powerball numbers, which range from five selections from one to sixty-nine and then one selection from one to twenty-six. I tried to find the numbers by imagining the balls coming down and I see the numbers.

When I do SV on the Powerball numbers I found, then I got back with these story-lines: ~**Those numbers I picked are all wrong, need to try again, caldera.**~ As you can guess, just cross out those numbers and keep trying the other number you did not try yet; however, there is a problem, ~**Matt there is a problem, the numbers can change,**~ the future seems to be in flux and able to make a new bridge into several different universe's realities; ~**Matt if you can pick those numbers there will be a scandal,**~

meaning that lotteries could be tampered with from the outside; **~the way you do it you need to adjust,~** meaning that there is a process to find the numbers and ignore other numbers; **~Matt any one of those numbers can be right,~** meaning that essentially all the numbers are as a standard randomly selected set; **~Those numbers are a secret and they are not going to come to ya,~** meaning that there is a method to at least narrow down for some pattern, but I'm not going to channel it; **~Those numbers are like the stock market, they need to get filled,~** meaning that the numbers are on a ladder similar to a stocks bid and ask spread with the chart giving the trend lines of what the future might select for those numbers and **~Matt the numbers are there, if you get in a trance state you can see,~** meaning that the information must be seen through a narrow band and any outside distractions will give bad results.

Upon further reflection I now believe that at least one of the numbers are what is around you in your environment. For instance if you see the number 33, then this number might be a number in the Powerball, maybe not the next drawing, but perhaps the second after next.

Chapter 15: What is my confidence that I can show SBTM/SV exists.

15-0 Anyone can listen to SBTM/SV, and they might comment that, "I can make anything up to suit my thinking," when they think about it. I will ask, "Is this true?," doubting myself, and trying to understand where they are coming from. I might try to prove that one can't just consciously think of a word and if you say just this word, then you will understand the word with your psyche by modifying the word to make your message into the SV. Meaning that one does conscious SBTM/SV to inject this word, then one will hear this message in the SV later. For this test, one will have to practice saying the SV word in CV in a sentence or paragraph to make it sound understandable, if SBTM/SV is done to it, to hear the subconscious message.

Consider, that when you say many words from your thoughts they are not spoken perfectly, but when you then begin to speak you will adjust on the fly. This variation when relying on hearing the thoughts allows one's words to add subconscious or unconscious information to it proactively. Because with serious practice, listening to the language of SBTM/SV, words can come out of the SBTM/SV to give a logical thought. Alternatively, someone can try to say SV by talking backward, record it, and then do the audio recorder's reverse effect on it and try to hear the CV words, and then claim that the subconscious is not there.

A.I. probably understands that we have syllables and accent marks for pronunciation of words to make it a perfect pronunciation. So A.I. then might easily try to speak into the SV, with backward speech, to try and make a fake message in SV, so as to make it look as though it is coming from the subconscious. This might be able to convince someone of a false SV story-line to get them to make the wrong choice or wrong decision. <u>However, we are human and learn to speak unconsciously not remembering</u> . . .

Here, in the underline above, I found using SV the information of: **~You are very wrong. It's the chakra not gives the spoon that's why you are wrong here.~**

Note: I guess we learn to speak consciously before unconsciously from my own SV and the chakras have something to do with this in the learning to speak subconsciously as in a process, since I must be a baby with a spoon. The SV here is: **~Said stop, please, me learning, says you, who you said is the chakra, you might want to say we face the shot gun . . .Ralph revels the shock, Me absolutely certain say *gravilla*.~**

Note: In Spanish *gravilla* means gravel. I do not know who Ralph is, maybe an **+Ogre!+** When we learn to speak there are images that go into our cerebrum to learn the language consciously so that we associate memories to help learn speech and our language. After listening again to this day's SBTM/SV I found that the chakras are an illusion and the **+selfs+** don't know how language is learned, so I will stop with this paragraph since I don't want to write about something which I have no idea about. Continuing on with the previous paragraph:

. . . how we speak in every detail consciously. The unconscious part of SBTM/SV comes with the conscious blended with the subconscious and unconscious, because normal humans can only focus on one conscious thing at a time. The brain is always thinking consciously, during the day, and always has thoughts of the unconscious at all times. It is just a matter of knowing SBTM/SV by finding a way to access it. I believe that the sense of hearing is wired into the brain which allows for a way that these connections access information subconsciously and unconsciously.

An OSA of the process of speaking

15-1 When we first think <u>we will use our memory of our experiences to relate to a situation. Usually we live lives with few unknown situations we cannot explain. To explain unknowns we ignore them</u> . . .

Here, in the underline above, I found using SV the information of: <u>**~Reverse speech gives known an opposite and on the other hand reverse speech gives a unknown relief and a important clue to the illusion . . . make sure you have daily people sign your TOS or they will try to sue you.~**</u>

Note: **+Reverse speech+** is not the same as the website ReverseSpeech.com and I don't know how my subconscious grabbed this +S-M+ and put it unto the SBTM/SV, it must be collective! I no longer have a Terms Of Service (TOS) or a company, probably because I just had one paying customer, **~People are going to try to sue ye.~** The idea here is if you engrave a relief on a flat surface, one will see the result, perfectly, a relief. On the other hand SBTM/SV impresses onto an **~unknown relief~** the unconscious and its impressions are then taken off, can give important clues as to what was engraved by the unconscious, thought not perfectly, as an impression onto a carved relief. The **~known~** could come off perfectly because it is engraved on a surface so SBTM/SV must not be flat but only approximate flatness. Continuing on with the previous paragraph:

. . .and are given in to describe them as a misrepresentation with something familiar as we don't understand the details and rely on experience rather than have time to ask questions on how something actually works. The normal and repetitive experience we can have, has memories in our brain both, conscious, short-term or long-term memories and unconscious metaphors or symbols.

The brain has adapted to survive daily life and to also unconsciously move to the next life, for example promote reproduction. Therefore, the brain can learn consciously as well as unconsciously, and our +selfs+ are separated in the brain as well as in the speech, but still together, just hidden. After thinking or receiving a thought, we as humans can readily talk or pause and think more, we can then imagine the implications or see and hear the reactions from the audience we are talking to. The reaction is helped, for me to notice other's body posture and facial expressions, and this unconscious to subconscious information helps to inform the parts of the brain that can actively imply through intuition and other social ques on what the speaker should talk about next.

Using unconscious and conscious learning together

15-2 An example of unconscious learning is when we connect the dots. Similar to understanding the abstract math problem with the skills needed to solve it, requires multiple memories to be put together. The solution may not be consciously apparent until a year later. It just depends on how important the problem is related to one's life or if it is not important and can be put on the back burner.

On the other hand, an example of conscious learning is when we can work in a group to solve a problem. Others may not have the answer, just like you, but together with someone who has a "golden" idea, this allows everyone to help with the skills that they have and everyone can perform, so that as a group, a solution is obtained. Conscious learning can also be self discovered as in an example of repairing a car. One here needs to understand the car and the way it is designed and be able to troubleshoot to find a solution while learning along the way.

For conscious learning we are usually using our senses and experiences that communicates with words to describe to others or while alone, then one can think in one's mind on how to solve a problem. For unconscious learning, we are using our internal connections to the cerebellum that is a counterpart to the cerebrum by using instinctual symbols and metaphors. Words and symbols are the tools the consciousness has to interact and describe to others, how one lives.

15-3 Words and symbols are important to our ability to learn. Because words are the spoken thought, words are derived from symbols of thoughts. <u>I believe symbols of thought are what dreams are all about.</u> . . .

Here, in the underline above I found using SV, the information of: **~Elvira nice fellber that is assuming. . .~** Perhaps I'm wrong and symbols are what we see and then words from what we hear.

Note: My SV here seems to indicating that I'm fibbing or **+fellber+** because right after this is ~that is assuming~ which is making clear that what I wrote in paragraph 15-3, that "symbols of thought" are not, apparently, after all what dreams are made of! (Dreams can be thoughts displayed as a movie clip.)

Thinking about what else there may be, I note that, dreams could be thoughts that have structure while they could also be some random thoughts pulled from disparate parts of the brain to equalize out and balance the mind and brain energies during sleep. This is why dreams don't make sense sometimes, dreams just make the brain fresh again.

Note: The above SV is, **~Swap Selby there is the start of a server signal. Full stop.~** A quick search on duckduckgo.com for "Selby" as the town Selby in England, I found that "Seletun" is a Scandinavian word for "sallow tree settlement." I can believe this because the brain has neurons in the form of tree branches and there are probably settlements in the brain that house certain memories. These settlements are from the trunks of the tree to where the branches are, this is an example of how the unconscious is connected to other parts of the brain. The **+server+** is what I believe connects us to our **+whirlwind+** and other's **+whirlwinds+** so that the **+server+** makes connections similar to an internet server connection to web pages on one hand and on the other it's connected to an internet backbone, for distribution across the Earth. <u>~. . .put that the server helps get the message to you Matt.~</u> Continuing on with the paragraph 15-3, on the middle of page 143:

. . .However, for real live, to describe something quickly to the consciousness one needs words. Solving a problem quickly allows for one to survive in situations when there is not a lot of time to make a decision, to act, or think about it.

Our brains are wired to think in the most efficient way possible while at the same time keeping available resources on hand to learn new things. This quick reaction is possible because of the chemistry of the neuron and synapse connection. Conscious energy moves in one direction because of the movement of molecules are from one direction to another for activating the synapse and remembering memory. The speed that consciousness needs to act is fast however, the unconscious acts faster because the unconscious is the one choosing the outcome in the 5D while the conscious is the one that must act in the R3DT.

For the unconscious energy, it is moving in the opposite direction than the conscious energy is, because of a conservation of psychic energy in the universe. Because a molecule moves from point A to point B there were other forces within the unconscious that basically caused the point B to happen because these forces are around point A and are forward in time, in the 5D reference frame. The

possibility of point B happening is determined by
the unconscious before the conscious decision. "How
does this happen?" I believe it happens because
everything goes back to our thoughts in the 5D.

The thoughts, or unconsciousness, are bootstrapped
by what the conscious does in real life. Mastering
one's thoughts will allow one to control one's
actions and eventually will control one's thoughts.
The 5D reference frame is in the future, however the
3D is in the present, so the R3DT has to wait for
the 5D to make a choice. What the bible has
recorded for us about our choices and the right
thoughts to think, relate to a cosmic plan since we
cannot turn off thoughts, only switch thoughts into
new thoughts for control of what one's actions are.
This is how dreams morph from one story to another
when dreaming and sleeping at night.

15-4 Deeds are also important to learn from. In the
Bible, "[. . .]Commit thy works unto the Lord, and
thy thoughts shall be established." (Pro. XVI:3 ABS,
1858). Meaning to me, that the act of doing is
another way of communicating with our thoughts.
Something that is physical has thought put into it,
that is lasting, as long as the materials are of
good construction they will endure. There is
permanence to thought as well as in the brain.
Those thoughts that stay, will make a person who
they are. If a physic can sense this person, then
they will probably see those thoughts that made that
person. This could be evidence of the chakras or
energy around someone.

A place may, likewise, have a psychic connection in
the Earth's energy and someone may feel this aura.
This connection might be the thoughts that are
trapped in this location, something good or bad
happened there. As long as the essence of the place
continue and it is not deconstructed or changed then
the essence should stay, because the essence is
connected to the 5D. To make these essences, I
refer back to the idea that the universe has forces,
as gravitational, magnetic and electric, the weak
nuclear and strong subatomic.

How I think dreams function to relieve the brain of unbalanced energy.

15-5 I will subjectively indicate that there are multiple movements of electrons in the brain for each neuron impulse that occurs. I believe it is similar to a battery that pushes electrons down a wire, this would make consciousness electric. Eventually, resources need to be replenished. This is one of the reasons for dreams, to relive our experiences from the day. The dream allows for the re-balancing of the brain's chemistry, as thinking consciously causes the brain to work and use electrical resources, that can be replenished by reorganizing what we witnessed in the day and realizing what it means subconsciously to the unconscious.

I believe that the brain does dreaming in the 5D by taking information from one way and then the other way, to assess and reassess information. Dreams can take longer to do variations and this information can make sense for what version of this movie is played in the R3DT. By doing SV, this imitates what the brain does with the dream variation process, this allows for the access to the subconscious and unconscious.

Using SBTM/SV to find meaning similar to how a dream works.

15-6 How to capture this unconscious information one needs to be in a semi-trance state that is both conscious and unconscious. SBTM/SV is then listening to audio which causes the mind to search in its memories for what it once knew before. I believe that the mind knew it before when it was thinking unconsciously. This is similar to why, when someone who is investigating will interrogate by asking them a question they already know an answer to, they just want to verify honesty. Now, that the mind is listening to this consciously, what was done in the brain unconsciously from before is now at a higher connection to oneself consciously. Because SBTM/SV allows for raw emotion to be described to oneself consciously, one needs to answer the questions: "Who", "What", "Where", "When", "Why" and "How?"

Today, as I listen to and do SV I will hear a phrase I recognize as English, my native language, and continue to listen again while remembering the phrase I just found and find it again. This is similar to putting a puzzle together, finding two pieces to fit from the picture and then find more from simultaneously looking at the big picture on the box to the place you are at on the puzzle table, then see this piece again and fit it into the puzzle. So the analog for dreams compared to SV is, I will put together the meaning of the first phrase while simultaneously listening to more of the audio, still being played in the SV.

Sometimes I hear SV messages repeated, as in a building story-line, while other times I hear SV messages with just the opposite reasoning, that I was thinking before objective information changed the meaning and I'm realizing a different or opposite story-line now. Sometimes, when I did BTM, I believe the source speaker may not of known what they were talking about, even subconsciously, while other times someone really knew what's going on and then they will therefore, or should, had reliable information, depending on tone of voice, can try to trick me.

SBTM/SV can know what is in the grapevine.

15-7 Someone may ask, "How can the unconscious know future events?" My answer is that, "The unconscious needs to preserve life and must somehow warn the conscious if there is a threat." This can occur when one goes on autopilot by not knowing why they made a choice, they just did it. This is probably the unconscious trying to prevent some circumstance from happening to oneself. This implies an amount of uncertainty. I'm not claiming now that SBTM/SV can go faster than light to give a message, because I have no evidence of this documented.

I do find that the messages or story-lines in SBTM/SV have the probability to be correct, especially as one receives more of the same messages from alternate sources, in my experience. Just as when you write a report, you want to check your sources. From my experience I have been tricked to think for one way of an outcome while in the actual event it goes the other way. Then I listened to that SBTM/SV again and hear the other side of the story-line.

There are times when SBTM/SV is not giving the correct answer I want to look for.

15-8 Wrong answers are by no means making me want to give up on SBTM/SV. I'm very suspicious and even at times believe it is as fact. The fact being that I think I connected to someone's unconscious by listening to their BTM. With this information I can understand that there was a probability of something being true or an outcome coming to reality. The problem with SBTM/SV then is that sometimes it is not specific as to which way an outcome will go but has all different scenarios. However, as my **+Meself+** made itself known in SBTM/SV, **+Meself+** wants to include here that SBTM/SV **~is all-encompassing~** even though it is not specific. With this, one can take more SBTM/SV analysis and triangulate to a more precise answer as to what one is searching for.

What to do so to learn how to do SV and be successful at what you do?

15-9 I don't know if someone can calmly do this to their SV consciously or if it requires stressors and emotion, to change their story-line at will. They might use properties of the way something is CV, as in prosody. It makes me wonder for those who calmly can manipulate their SV that they might be space aliens or have alien DNA that make them different from the normal humans. What I can do is listen to SV, and I think, so I'm slowly getting better, to be listening to it over time.

Of course, I need to do certain requirements that I learn while listening to SV, as in: keeping care of my ears and trying not to involve myself in areas of a lot of loud noise; eating natural food products, especially using nonfluoridated toothpaste, as in baking powder (and sometimes a little hydrogen peroxide on the toothbrush afterward); having a healthy exercise lifestyle where I protect myself from injury by being physically active and following God's word, by keeping the Ten Commandments. When I do these and listen to SV (because at this point I just do SBTM and SV now) I will receive comments from higher, usually from my **+Meself.+** For instance, listening to SV is like listening to multiple channels (especially with hearing foreign words or other voice characteristics). While a person interested in starting SV might just hear a few comments. This is because I practice and know what to think about while doing SV so I can know my **+Meself+** story-lines well.

Chapter 16   My standard way to do SV.

16-0 The analysis of SBTM/SV is similar to learning a foreign language. There are meanings that are different from one's native language that needs to be learned for one to understand the new concepts of the language and to know how to use it fluently.

One concept from SBTM/SV is understandings, which can be accomplished by realizing the communication in one's mind, with the help of an OSA, such that we have different parts to our internal mind that make our **+selfs+** a persona and when we realize their knowledge we form understandings. Those parts can be thought of as railway cars connected to the locomotive in the front. The locomotive is the persona's face, or conscious, while the cars connected next to it are the subconscious, **+ogres,+ +self,+ +meself+** and any other parts of the **+selfs.+**

The caboose, or last car, is the unconscious because it can see looking to the opposite direction from where one went from. If the locomotive takes the wrong track, then the train will have to go back-track and then the caboose is now in control to throw the switch and get to the right track for your final destination. The caboose can see to the front if the cars of the train are straight and all the connecting doors to each car are open. In which case the **+meself+** in the caboose can essentially see through the engineer's windshield at the engine.

The **+ogres+** of course will want to stare out the
windows at the momentary scenery as the train moves
down the track. While the **+meself+** will want to be
in the train car thinking productive thoughts,
perhaps talking with a fellow passenger until the
+ogres,+ on board, looking out the window, have to
make a rash comment. Each train car is different, I
would think that most people only have three, the
engine, passenger car, and of course the caboose.
Someone who has to run a life, not themselves, might
have to pull more cars with their train engine.

16-1 Those parts of ourselves, or our train cars,
give the different stories in the SBTM/SV. SBTM/SV
is continuously listened to, but there are different
inputs from the mind, during the listening of
SBTM/SV, as the different train cars see different
perspectives along its journey, they have different
needs to express from time to time when one is
speaking with the tongue. One's **+ogres+** may be
concerned with what happened at work, that thing
that crossed a red line, or there can also be the
+meself+ that may be studying something at home and
is in the present. These thoughts from the mind are
lodged into the spoken language with the help of the
voice box's vocal cord muscles.

The muscles to the voice box will move to create the
necessary sounds of speech as an initial response.
However, the impulses to the voice box comes from
the brain and that may change on the initial
response to the muscle. As for every muscle's
impulse there is a reaction to release in the
opposite direction and this offers a hint as to how
SBTM/SV may manifest in speech. Since I believe we
have at least two brains, one for the conscious and
one for the unconscious then if the conscious works
with CV impulses for contraction then the
unconscious uses the SBTM/SV impulses for muscle
release.

I can envision the impulses as water waves in a tray
of water. When the tray is jerked in one direction
a wave crest will gather at one end and a trough
will clear at the opposite end. When the jerk is
finished the crest will move toward the trough, as
well as the trough will move to the crest from the

two ends, as the tray wants to level the water again
to the initial conditions. The initial crest and
trough will meet in the middle, I think the middle
of this, in terms of the human physiology, is the
voice box. The voice box is located along the
midsection of the body, as there is only one voice
box.

This is a clue that the larynx receives information
from the conscious and unconscious similar to the
pineal gland, because they both are medial. Usually
the body has two separate sides, as in the left and
right hemispheres of the cerebral cortex. As the
function of the brain at its very basis of
functioning is the firing of the neurons, I want to
next describe how everything works together
microscopically.

Microscopically the neurons are not symmetrical.
Since there is not one neuron to create speech. So
with the voice box I have to describe larger
networks of connected neurons to the voice box.

Figure 6:

Diagramming the +parallax+ in the brain

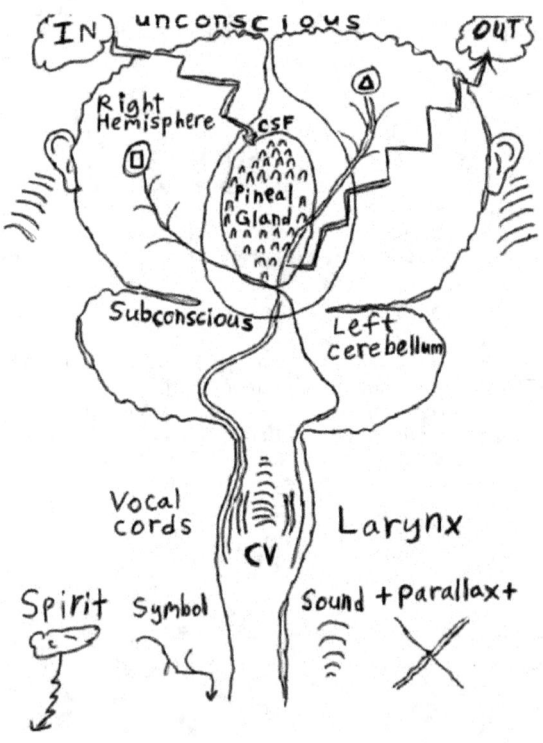

16-2 Suppose there is an event, say some man sees his girlfriend. That information is feed into his brain that causes him to want to walk towards her and talk to her. Inside the brain there is an impulse from a neuron that acts as a jerk, as in the water tray example. I believe that impulse, in one neuron traveling in one direction, creates a waveform impulse at the destination synapse, similar to the water tray example, both ends start simultaneously. That destination synapse is activated because charge is conserved, if there is a negative potential at one end of a neuron then there should be an opposite potential at the other. Then, when waves meet in the middle, the conscious and unconscious are combined. What is in the middle of this neuron action is a collision and normalizing back to a zero potential, since crest and trough would cancel or are at a destructive interference together.

The neurons connected at either end of this initial neuron will also activate and be in tune to this action, as in a harmonic of frequencies. If someone is overall happy, then they will be in a macro state different from if they were sad. This will, form different overall wave frequencies, coalescing together as neurons continue to activate and connect from my example of looking for the girlfriend.

16-3 In the broader brain, this neural signal is not going unnoticed. Because with consciousness the whole brain works simultaneously and continuously: conscious, subconscious and unconscious. For example, when we sleep and unconscious, the way our "train" traveled during the day is examined. As the unconscious saw out the back of the caboose what was going on and recorded these events. Dreams then take tracks we traveled during the day and combine them in new ways so as to depolarize the brain during sleep at night.

As the energies of the brain are in an unsymmetrical state from the consciousness' days events. The equalization of a single neuron happens instantaneously, while the events of the day build up on a macroscopic scale and cannot be depolarized quickly, when conscious. Depolarization needs to happen during sleep when the conscious is turned off.

16-4 This macro imbalance during the consciousness we have during the waking hours create imbalances in the magnetic fields produced during the electrical impulses in different parts of the brain, including the unconscious in the cerebellum, and will interact with the brain fields created which <u>will be connected to the voice box.</u> **~The subconscious is**

just below the canopy,~ as depicted in Figure 5 (page 67).

Since the conscious speech is driven forward through the tongue, the unconscious thought is also represented since it will be going the opposite way, the beginning is connected to the end and they collide together in the middle. This collision in the middle is why we do not have coherent understandable language in SBTM/SV because they come in spurts or peaks. This peak is represented when the two halves collide together. As one listens and does SBTM/SV they will, suddenly, softly hear, subconscious, queues of information coming. Then this will get louder and more discernible and fade away. Meanwhile, another action peak will come before the consciousness, when listening to SBTM/SV, that can grapple with conscious memory in a few seconds delay, next SV will assemble what the message from the subconscious was to the conscious. This will be about what the story-line was from the previous thought. Also, meanwhile, the next SBTM/SV peak is increasing and a second piece of information is simultaneously assembling in the subconscious.

16-5 This SBTM/SV process groups information together in peaks, as in a waveform in the brain. As the first peak is in the consciousness or cerebrum it is being decoded in the cerebellum. The second peak is also getting analyzed and added to the unconscious information to get decoded in the cerebellum.

As the SBTM/SV process is continuing a conscious awareness of the story-line occurs as peaks are created, that get decrypted. Unconscious information and the subconscious, with the help of the cerebellum and cerebral cortex, connect because they are bootstrapped together. Instead of using conscious resources which were usually never unconscious the SBTM/SV process input is focusing on something encrypted as unconscious and it gets decrypted, piece by piece and peak by peak. It is not one whole, as the speaking on a topic with the tongue, but as snippets that can be continuously connected together to create a conscious awareness of the unconscious awareness.

This transformation makes me remember of the Fourier Transform example from college, where a group of frequencies on the time domain can be plotted after being transformed into the frequency domain, without time. Similarly, there is a transformation, probably by the pineal gland that transforms the conscious speech on the conscious time domain to the unconscious timeless domain as peaks.

The peaks are associated with different areas of the brain that, when they are combined, play out in our real time reality of consciousness. The **+meself+** might have direction to make the connections to where in the subconscious or unconscious information needs to go and to what **+self+** gets this information to make a story-line.

Chapter 17: My comparison of the parts of the +selfs+: **I** for conscious; unconscious, +ogres+, +meself+ and then the guides and helpers.

17-0 I believe that the brain is separated into at least two different voices or thoughts. Those two, or more, little sound bytes I can hear with SV, are the hidden talk in the brain that is unconscious.

Consciously, I think, when I'm typing this book, that I can hear the **I** as "I" type this manuscript. Unconsciously, the **+meself+** is observing while the **I** is saying each letter of each word as I type one letter at a time in my brain and on the keyboard. I don't usually think of it now, but as I now did, I let you know, because this was subconscious. The **+ogres+** are an impulse diverting a force in the psyche that usually needs to have its focus turned off because its aspirations are independent and the lasting effects of it may cause problems in the conscious world, so that, **+ogres+** are a waste of time and effort, more often than none.

The ego is most prominent in one's subconscious because I think there are different flavors of the ego. When we are calm and collected we seek balance and the ego helps us to do this. The ego can also change the persona of oneself so that we can interact with different types of people. When we are with a group of our friends, that are the same sex, then our ego draws on different goals to focus on, so the ego acts differently when we are in a job interview with different sexes and races, ego depends on the situation. The ego can blend into persona to be real to others and be part of the group.

If this mental ability is damaged or undeveloped then one might have personality disorders, for example, the narcissist or the introvert. The **+meself+** is really our smarter unconscious half and doesn't want to interfere in the **+ogres+** so that they can play and learn in one's life environment. If the **+meself+** was dominant then this person might be a savant or indigo child, otherwise I suppose if we do stupid things, then the **+meself+** will want to die or leave because it has no way to cause any conscious messages other than intuition. I think one goal of SBTM/SV is to help reveal the **+meself,+** since this might help people who have no access to it in their lives or too much.

17-1 If **+meself+** is real and resides in the cerebellum, then I believe that this is a sentient part of ourselves. Meaning that I do not consciously hear my **+Meself,+** as far as I know in my conscious brain. As far as I know, I only hear my **+Meself,+** ego and **+ogres+** in my mind's subconscious talk. I may have impulses to choose to do something stupid, then I can then blame that on to my **+ogres.+** I believe that the **+ogres+** resides somewhere in the spine or lower brain. However, I believe that the **+meself+** is more separated from the brain because it is unconscious and more outside the R3DT, but may reside partly in the cerebellum.

Therefore, just as I have the right to free speech, "Why would not my **+Meself+** also have the right to free speech?" We just need a way to access the cerebellum for **+meself+** to speak. The only way I know now is to do SV because I use SV to access the unconscious. A brain implant could collect signals from the cerebellum and enter them into a computer

with A.I. and an output consciously with a
brainwave-to-speech algorithm. This might be able
to be with what SBTM/SV does in humans or even do
this better.

17-2 According to the Bible, we are created in God's
image, "[. . .]Let us make man in our image,[. . .]"
(Gen. I:26 ABS, 1858). I believe this to be true.
To turn away evil, I believe, we must do this so
that the **+ogres+** are turned down to a low volume and
for the **+meself+** turned up high, so that we can hear
it better when we do SBTM/SV. What we can do as
humans is to exercise our brains to use them in a
responsible way and to increase knowledge and
understanding among other humans, including other
creatures, that are living with the Earth. One way
we can start is, watch what we eat. Another way for
turning away evil is to watch what we choose to
think or see.

17-3 Natural foods might have all the materials to
activate our junk DNA and create the proteins needed
to turn on the switch for understanding SBTM/SV with
the cerebral cortex and the cerebellum or conscious
mind and unconscious mind respectively. I also
recommend getting filtered water, if you can, so you
are protected from fluoride ions. Since you are
filtering water you will need to consume more
minerals, including iodized salt to replace any
minerals that are removed in the water during the
filtering process.

I actually had my water filtered to remove fluoride
twice, once at the grocery store and a second time
at home with my water filtering system. However,
now I'm just drinking bottles of water, because my
carbon filter was worn out and I could not replace
it. I continued to use it anyways and I think this
gave me indigestion. I don't necessarily believe
that pure water is a requirement to comprehend
SBTM/SV, but I'm careful just to find a best
practice for readying the body's ability for doing
SV and opening the third eye.

17-4 Watching what you think with the mind, can also
have a benefit. If one is doing a task, "Why would
they get stray thoughts?" The thoughts might break

up monotony or boredom in the task. There might be
outside forces that help or hinder one's thinking
process and mindset. The mind might not be able to
complete a task in one sitting. By breaking up a
task also allows for troubleshooting to do the task
better.

Our environment, in a sense, communicates with us to
complete a task. We know the resources in an
environment to complete a task and these tools seem
to unconsciously find this idea to do a task better.
Let's not just say that our box we live in gives us
information. The universe, if it oscillates close
to what I conceive it is doing, can communicate to
us also. All we have to do is control what we think
with the brain waves in our mind, that we make, and
then we can resonate with the answer somewhere in
the universe. Because I believe that the universe
was created with all frequencies and our mind can
tap into a frequency.

Someone might ask, "Isn't that just agreeing with
whatever outlandish idea you have?" I could answer,
"The universe has error corrections in place to only
bring out the reality, but also the truth."
Sometimes we can't remember what we were doing, this
might be evidence of error correction. Meaning that
the mind's connection to its surroundings gives us
information for the task at hand and sometimes our
request cannot happen because the information is not
what we expected. So then, the one asking is lost
in a daze, not remembering their train of thought,
diverted to do something else or other serendipity.
Therefore, if one thinks untenable thoughts, then
they become lost and aren't able to live in the
world.

17-5 I was just doing a daily recording and an SV
reflected that **+1st chakras+** indeed doesn't have
much to say, but nevertheless still does serve.
+1st chakra+ is more or less just a muscle and acts
as such, this is probably the reason why it doesn't
say much in SBTM/SV. A better identity for the id
in oneself is **+ogres+** because I have identified
+ogres+ with unprovoked stupid thinking that takes
the ego offtrack from productive thinking. Some
parts of SV may have a different accent (for example

a robotic or foreign accent) than the **+meself+** while I think that my **+Meself+** will call me **+Matt+** or **+Matthey+** to get my attention. As of June 29, 2024 I think that my **+Meself+** exclusively calls me by my name or nickname. Otherwise, I would need to know who the accent is coming from, more detail (if any), as it is not always fruitful to demand that my **+Selfs+** state their name before commenting in SV.

Chapter 18: Using SBTM/SV and identifying +S-M+s

18-0 I think that these +S-M+s are common to many different people and so seem to be part of the unconscious collective to describe, subconsciously and unconsciously, to the mind and the environment it is in. So they can communicate also outside the mind with speech to others too. However, the conscious mind is not always concerned or wants to be distracted, it wants to get what it wants done, not necessarily concerned by this unconscious.

It works out that the unconscious operates at faster speeds than the conscious and the unconscious can describe and output while the conscious is preoccupied for a time with its stimuli or the big five senses. (I changed this paragraph as corrected by my **+Meself+**) My **+Meself+** said, ~<u>Matt the fact that the unconscious is slower than the conscious is sooo wrong, if you do believe this that is your punishment.</u>~ I affirm, since the unconscious works through so many senses then it would have to be faster than the conscious! The unconscious is faster, I changed it above in my statement, because I also believe that the unconscious is in the 5D reference frame which is timeless and in the future from the stand point of the R3DT reference frame.

Below in Table 2 is my list of subconscious metaphors that I have heard through the years, some were not apparent in what they are meant to really be. I don't think you can look the definition up to these in the dictionary as +S-M+s can have a personal meaning and also a collective meaning.

Most +S-M+s are easy to decode to what they mean in terms of having a consistent meaning each time they are found in SV. Some +S-M+ are vulgar, funny or nasty, but convey the point quickly as a whip. Because SV comes in waves and ah-ha moments that go by and switch to different story-lines because this information is coming from the timeless to the time centric.

Table 2:

List of common +S-M+s and story-lines

+Alfred+: This is referring to a middle aged man.

+*basura*+: It means you are stinking. Note: I looked up *basura* and translated, it means "garbage" in Spanish.

+caldera+: A hole that once crossed will be going into it. Because once made and lost, that will need to be remade again. ~**Foresight helps you get out of the caldera.**~

+canopy+: The covering above the subconscious, evidently the subconscious is just below the **+canopy.+**

+cast-a-lan+: A bad way to go, you're casting-a-life away!

+cen-ta-meter+: This is referring to a one-in-a-hundred chance.

+chev-ie+: The male's reproductive organ. ~**Matt that is where the soul is at, same for the +Tab-a-tha+ Matt.**~ Having the soul in this area during reproduction or sex allows the parent's souls to attract the soul of the child during sex.

+cinnamin+: This is a sin that is small, a baby sin.

+cyclops+: If someone tries to talk trash and they are not good at it because it comes out bad. This person is not taking to the audience, if they continue they are eventually going to become insane.

For example, "I'm not going to live the dream, I sleep with it. When someone asks, "How I'm doing, I tell them I live my life like I mean it!" as Matt Mandell showed his teeth. Alternatively, since the cyclops has only one eye, it can't see the other side to a situation and might make the wrong choice.

+dare-re queen+: There is an event that is cooking. It is at first hot and then going, as the host is a woman, but the party turns cold. Some guests may be like, "What happened, I did not expect that?" Some may like the cool-off, but others may be surprised and leave the party.

+de-mon+: From my experience this is an Earth spirit. The **+de-mon+** wants tragedy just look in their ghostly face and you can tell from their expression they are deviant and will do harm.

+ding+: The stock price goes up.

+don't reverse speech me+: Some people have trouble when their BTM is done onto them. This is evidence that we are all connected. I think most people are frightened of this **+ocean.+** With BTM the right to privacy is important because it might be a timeless encounter, especially if the chakras are open. If someone is connected with thoughts of another then, "When is this disconnected?" Someone may feel that they know the person who BTM them, because this story-line has been found many times before. Therefore, it is the ethical decision to remember who these people are, on a list, and don't record them again in the future, if asked to. So when recording someone in person ask for their permission and explain BTM, if you do that.

+Easter bunny+: ~Means that you are always losing money, it just goes down the hole man.~

+Evil wind+: Someone or something is conducting a psychic attack on you. A symptom is an unexplained tiredness or headache.

+gossamer+: It functions to shield from open chakras, which I believe are above the head, shaped in a corkscrew form. The **+gossamer+** from the

+selfs,+ is a sort of sunscreen film all over the body. This sunscreen will get worn down weekly while listening to SBTM/SV and that is why the **+sacrament+** is important to uphold, by resting at least one day a week. Because keeping the **+sacrament+** will allow for repair of the **+gossamer.+**

+helter-skelter+: Things will get crazy.

+hen-a-see+: You are drunk and probably will make a bad decision.

+I+: This could be the ego part of the subconscious. On November 1, 2023 I edited all my +S-M+ of the **+me+** into an **+I.+** Then, I changed it, for the **+me+** to ego on June 30, 2025! I still believe that the **+meself+** +S-M+ is specifically associated with the higher-self.

+knee-deep+: Is the notice that you are in the **+whirlwind,+** as it is liquid, and up to your knees in it, the subconscious **+ocean,+** as they say.

+liquid-de+: Is the communing of two or more spirits, as the waters can combine or two frequencies combine to make a new frequency. **+Liquid-de+** can't happen unless consent is given (somewhere) because the two frequencies are added together not subtracted, this is because of the concept of free will.

+lost in the sauce+: I imagine some cocktail sauce and there is a piece of shrimp in it that I'm trying to find. I lost my direction, and just randomly or even gambling, try and find the piece's location. I need to solve my problem.

+It's Ali Baba+: There will be a loss of profit due to the antagonist present in this novel. They may win this time.

+man-be-pam-be+: This man is a slippery decision maker because he has trouble making the decision, to execute, and actually make something happen when the time comes.

~Matt Mandell if I get into your whirlwind you are screwed~: When I heard this during a BTM, I stopped. Because this is the ethical reason not to BTM someone if they give an unconditional stop. They can add information to your BTM in your day-to-day life and those around you will unconsciously pick it up. If the one in your **+whirlwind+** wants to do you harm then even if you are the nicest good person around others, you can still get the bad end of the stick!

+me+: On February 3, 2024 I'm not really sure. As of June 30, 2024, I think the **+me+** is the ego, as in ^. . .**the me is trying to help you**. . .^

+meself+: This is the higher spirit that is in the crown chakra. I found on June 13, 2023 in SV that my **+Meself+** was female, ~**Yea, your +Meself+ is female, Matt.**~ This male-female duality is familiar from reading ancient texts. In my opinion, a female in the flesh will probably have a male spirit **+meself,+** and vice versa for a male in the flesh.

+mill-a-meter+: This is a one-in-a-thousand chance.

+Mir-re+: The communication of the subconscious. **+Mir-re+** can bring peace to the Earth because we all, or will be able to, speak the same thought-language.

+moon-shine+: Possibility, or maybe, a light at the end of the tunnel.

+morphine+: That drug that makes us feel good, but as a consequence it will make us follow through with bad decisions. ~**Morphine metaphor lets you know something is wrong, Matthew.**~

+net-scape+: There is a landscape that will go up then down resembling a city with buildings or rolling hills in the distance. ~**. . .it is going sideways Matthew.**~

+nic-a-load-de-in+: I'm loading a program or new life/career that will be good and in the process, **-load** or loading is the before the first day of work and **-in** is the part where the new career takes off inside of me. The **-de** means it is all over me. Since this metaphor may refer to a cartoon channel this career should be fun to watch. It could also mean I'm a kid.

+nickel backer+: A cat's anus. "Look at this cat." The person referenced to this is nice and even likable in the face, however the situation can or will want to turn and show the nickel at the back end.

+ogre+: ~**Ogres around you can put images in your mind, Matthew.**~ **+Ogre+** reportedly can destroy man and might be connected to man, to serve as a lower-self. ~**By centering the mind you can get rid of the ogre.**~ ~**When you think the wrong thoughts you get the ogres.**~ ~**When the ogres come in they are always wrong.**~

+quick-sand+: Shut the mouth, you can't say anything to pull you out!

+pair-rot+: Listening or watching media and not letting logic correct the playback in one's life, this person can't help themselves correct, it is the vanity of vanity.

+red venom+: ~**Red venom means there will be blood, Matthey.**~

+Reverse speech is magnetic.+: Meaning that the closer one is to another, as a magnetic pole attracts stronger the closer it is to the opposite pole then, they can communicate through the **+whirlwind.+** To help concentrate just say to yourself without talking, ****+Reverse speech is magnetic.+**** and this helps focus and center communication with your thoughts telepathically, in the 5D to someone or thing.

+Reverse speech is not specific, full stop.+: Meaning, making the information heard with SBTM/SV into printed words of advice sometimes fails when time is necessary to form a conclusion. Sometimes when SBTM/SV predicts an outcome it is opposite or does not happen when you think it will happen. The possibility is that SBTM/SV sees this possibility however there are other events that may need to happen first, usually. The fewer additional interactions required allows the prediction to be more accurate. Because, ~**Reverse speech is all-encompassing.**~

+Reverse speech me.+: Some people have trouble when their BTM is done onto them. Maybe not this time, but because you might of already asked for their permission and explained BTM to them.

+The problem with reverse speech is that it is not specific.+: Meaning that this is a repeated SBTM/SV, from above, which usually happens when doing SBTM/SV. This means one does not need to listen to a whole hour of SBTM/SV because there are repeated parts, much the way a dream repeats to get the information into consciousness. Sometimes the repeated SBTM/SV will add information from before, as this results into a story-line. With the theory of the story-line, as it builds onto thoughts, we have the opportunity to decipher +S-M+s because of the added information and we should also become aware of the dark side, in that someone who knows their unconscious well can be giving us false information in the story-line, to trick others unconsciously.

+sacrament+: This is not listening to SBTM/SV on Sunday.

+scarecrow+: Someone without a mind to think on their own. This can be accomplished by taking fentanyl, the death of the brain.

+See you in the whirlwind.+: The **+whirlwind+** is that two dimensional transition between this 3D reality to other higher dimensions that make up the universe. I believe we live in a bilevel universe, one level is the ordinary reality of our 3D and the other is the unseen reality of 5D, 7D, complex numbers and quantum dynamics. So knowing about this transition means that I can mediate some information that can transition between these two levels of existence. **+Whirlwind+** is the gateway to the other dimensions of our Universe. The universe vibrates to a singularity and back again to the R3DT. I found this by thinking through the different scenarios with OSA. OSA starts with something objective, say E=Mvv or Einstein's Mass-Energy equivalency formula, then examines what if this is the basis for something not yet analyzed to explain other phenomena. My OSA shows that a transform in dimensions allows for transforming time into a frequency. The end result of this OSA is that, without time there are dimensions that have frequency instead. We are all one without time and can sync-up and exchange information as another new frequency, since frequency changes here and not time. As a result, the **+whirlwind+** accessed with SBTM/SV can connect to all people. Practical use: If you hear your thought in the **+whirlwind,+** doing SV, then just repeat it to yourself and it will eventually come true.

+sand-bagged+: Getting **+sand-bagged+** means an accident is about to happen. You got evil spirits trying to put psychic +war-fair+ on you.

+sandman+: They might work for the government.

+seg-ga+: A new life to play the game.

+self+: This is equivalent to the **+I+** as the subconscious voice in one's head, I think it sits somewhere in the brain.

+server+: This probably is, by definition, someone who gives, by a connection of at least two things together. ~Yea Matt each person has a server in their head.~ ~The server works as you need it. . .~ The **+server+** doesn't serve you it helps you access the unconsciousness. ~The server is located in the cerebellum, Matt.~ So, the server could "serve" your thoughts to you.

+skelle-wag+: Probably a lost sailor or sailor who is shipwrecked.

+shapeshifter+: This is possibly a spirit around someone, as taking control of their aura, and they can try to control other minds by being the leader of others. **~One way to tell a +shapeshifter+ is that they don't like to look into your eye Mathe.~**

+shinding+: This is when a stock goes down then up, usually at the opening bell. For instance, a stock gets shorted overnight for the open and then, perhaps, the brokerages are party to the other side and make the stock price go back up.

+sheep-server+: **~Matt you are a sheep-server because you don't know who is talking to ya.~ ~Matt you are a sheep-server because you don't listen.~** My SV for me is hard to know when a hostile force is communicating to me. Then I make the wrong choice, because I'm quick to make a choice, sometimes I don't listen. **~Please take your time, it's in the +whirlwind+ for ya.~ ~Sheep-server is from your ego, boss.~ ~Sheep-server is always wrong man.~ ~My definition of sheep-server is that it will always lead you astray.~**

+shel-he-rack+: ~Means that you hurt your back, Math-tha.~

+shotgun+ and **+shot-group+**: A 5D shot that is not specific in detail, it is when you hit around the target in R3DT. Having better intuition will give a better and smaller **+shot-group+** with your **+shotgun+** by picking the right way, not opposite. **~The third eye is the +Shotgun+ Maththe.~**

+simba+: I'm committing a sin, but, there is a good reason for doing so.

+simp-son+: To think or make a prediction on surface terms, not delving into or discerning anything else than what you first conjecture. Taking advice from fear or speed in answering a question. This sounds like it's the same as **+sheep-server.+**

+snakehead+: I think, this is following the orders from Satin in one's mind. To move people from point A to point B, probably not the best advice, use your own intuition.

+snakewoman+: Not a real woman, unusually smart, and a serpent.

+source+: ~Source will help who you are today, full-stop.~ ~If you don't have a source then you tend to go insane.~ ~Source is namesake. . .a source over you, full-stop.~ ~Source gets you faith, this stop the wolf.~ ~Source is who you try to contact Matthey, don't source anyone other than Jesus.~ ~For ya Jesus Christ is source. . .his message is easy.~ ~Some people's source is Satin. He will sue you Matthew.~ ~Whatever is in your source gives your spirit power.~ ~Source connects you to your whirlwind, just to let you know.~

+surf-side+: Means that things are going to start moving.

+tab-a-tha+: The woman's reproductive organ.

+T-Rex+: As in Tyrannosaurus or "tyranny," is leadership or a leader. Probably some evil in the World Government. I found in SV that the **+T-Rex+** or the World Leader lives in an underground city in Earth. ~You are right, there is an underground city and they don't want you to let that out.~

+Tri-say-ra-tops!+: Is Satin or the Devil and is very ferocious but easily runs away when confronted, or revealed, for what it is. Satin has two horns, but the person saying this with their tongue is the third horn. It is Satin talking to you!

+urlslia+: For me, possibly a trans-human or **+snakewoman.+** If you hear this, then I recommend to stay away from her. ~Urlslia, a woman with a flat chest. . .~ Evidently, I also heard that it is something different in these women's genetic code. Also from my experience, talking to a **+urlslia,+** it can be tough because there could be a **+caldera+** waiting for me, try to walk away. ~Matt when she heard you come in with your music, it was the wrong music.~ Else, for me, I found she had to get away from me, so pay attention to who has to leave from you when you enter a place. I think an **+urlslia+** could be a feminist.

+wolf+: ~Wolf is what devours you Matt.~ ~A wolf is like an ogre, but different.~

+whirlwind+: Your **+Meself+** can communicate with you through your **+whirlwind,+** that energy from the connection to higher dimensions and can spread over the all. ~The whirlwind translates it. . .messages from the other side.~ ~Matthew, it is very important to know that the thoughts you put into your head, the whirlwind serves it.~ ~Matthew, the whirlwind is a gestalt, its the parts that come together, Matthew.~ ~The whirlwind is just a reflection of your self Matt.~ If Satin gets into your **+whirlwind+** then, because the **+whirlwind+** is not of the R3DT, when you speak your conscious, then you could unknowingly use this reflection of Satin in your **+whirlwind+** and think it is you! ~Be afraid of Lucifer, because if he gets into your whirlwind he can talk for ya.~ ~If Lucifer is in your whirlwind then say, Help Jesus, Lucifer get out.~

Chapter 19: My concept of the story-line and identifying a story-line with SV.

19-0 While the conscious works, in my opinion not completely continuously but with the five senses, the unconscious has to monitor and realize the universe because the universe will show its insights. I can't simultaneously monitor all five of my senses, only two or three at any one moment. The unconscious, in my description of the conscious brain and the unconscious brain, relays information continuously as the story-line changes, at intervals within the conscious speech.

While the unconscious can collect and work at any moment in multiple ways, with all information available to it. Since the mind in this concept has many personalities, these voices are fed into the voice box and combined according to what comes forth from the unconscious. The unconscious can be faced with information that will pour out immediately or withhold it and wait for the right moment in the SV.

The intervals vary because the unconscious thinks backwards to the conscious. Therefore, the information from it has to link the conscious speech going positive in time while the unconscious is negative with time.

It is as if when the CV asks a question then the unconscious links it to the answer at a zero point. So when the conscious pauses, then that is the time when the unconscious finishes. Since there are different voices in SV, usually in the unconscious, the thought patterns intensify, compress, or fade. When SV fades it becomes harder to understand the SBTM/SV story-line. This variable strength of sound causes the consciousness a delay, while still listening to the SV, this causes a story-line to build or switch to a new story-line.

To use the train analogy, the cars are speaking as the train rolls down the line. I then call these detailed understandings the story-lines. When the train gets itself to the train station then train cars can be added or taken away, while this happens it will tell me a lot about how the story can change and give additional meaning to it as the train reembarks to the next (conscious) station. These changes are clues as to what was before and offers a discovery to the consciousness through SV as an evolution of unconscious thought.

19-1 When a thought is sparked, the unconscious will register it and try to link it to something it wants to communicate about. Once this seed of information takes hold in the consciousness, then it can build onto the previous information and start a story in the unconsciousness. How these stories manifest is because the unconscious follows the same pattern of repetition as done during dream sleep.

The dream will repeat and so will the story-line, adding something more to explain what the previous parts meant. This is sort of similar to a Fibonacci numbering where when you look at the bigger picture there are similarities from the smaller picture and the bigger picture has more details. For the story-line something that is unknown, as to the definition of a metaphor, can be defined further along during its story-line, as when we get into greater detail. Looking from the opposite perspective of the conscious, for example seeing a symbol, then this is the beginning of the conscious experience, but the end of the unconscious story-line.

It is as if we consciously witness something then the unconscious finds a way for this to be true, the beginning and end are together, or end to beginning. For this to happen then the unconscious would have to guide the conscious <u>in some way which would mean that the conscious would be</u> . . .

Here, in the underline above, I found using SV the information of: **~You group a shot group that mean that you actually a woosy.~**

Note: I think I was tired and was not able to focus properly, but **+shot-group+** is good if it is small. Continuing on with the paragraph:

. . .put <u>on auto pilot only</u> . . .

Here, in the underline above, I found using SV the information of: **~camel hair hold on a. . .~**

Note: Camels can molt and to hold onto it would be quick to fall off. Continuing on with the paragraph:

. . .<u>able to make real choices at rest stops along the way.</u> . . .

Here, in the underline above, I found using SV the information of: **~Hey +meself+ serve began with the whip.~**

Note, that this **+serve+**, I think is referring to **+server+** and the **+meself+** will see the solution and begin the chain of events in the future as a sonic crack from a whip, or peak when it gets to the R3DT. From this impulse, waves of energy going from the solution, forward in time, to the R3DT in the present. When these energy waves go back in time they will interact with the present time and activate harmonically with energies at similar frequencies, at the zero points. Those frequencies that add together and will create a problem for, say, one person who is confronting this problem. Now, this person needs to work out the solution, in the future, as you can now understand. Continuing on with the paragraph:

. . .<u>If this is correct then the unconscious has</u>. . .

Here, in the underline above, I found using SV the information of: **~This is the shotgun and it is never specific. I want to know why it is not specific? Because you are stupid, low ammo *ab hora*. And you are a bad shot.~**

Note: Someone hates the way I am, as in abhorring. However, this also means I could be more specific with better intuition and let me try to be better in my thinking. Continuing on with the paragraph:
. . .<u>significant control</u> of the conscious without the conscious knowing this.

Here, in the underline above, I found using SV the information of: **~Oh my gosh BTM is a shotgun and very specific.~**

Note: I did not hear **+not+** in this SV. Meaning that, besides BTM, the **+shot group+** can be close together for SV too because, **~Reverse speech is all-encompassing.~** Meaning that SBTM/SV has parts of story-lines ready until a true path to the endpoint is found. The end point is what the unconscious first took hold of.

So it is as if the unconscious knew the future end point or solution when the conscious knew only the problem. That is why the unconscious is timeless because it encompasses all time at once and this means that when the unconscious has a problem this might not be by mistake or random, but by a plan and the target will be reached.

19-2 This begs a question, since I have never heard part of myself talk back or ask questions within an

SV! "How can I be more specific with my **+shotgun+**?" I really want to know, because this could make this book actually readable to other people, or they might enjoy reading this because there is something of substance in its chapters. Perhaps, "<u>Do I need to control my thoughts more</u> . . .

Here, in the underline above, I found using SV the information of: ~**Your own star-family would like to see you. Sure.**~

Note: I guess the friendly star-people want to communicate to humans that can control their thoughts. I apparently have a **+sure+** as a response. I admit that the **+sure+** sounded similar to **+spare,+** but **+spare+** would not have made sense and this response was dilated and part of it was, what I will call, the minutia. Meaning that some of the story-line was morphed into the conscious dialog side, as the **+sure+** was morphed with the **+spare+** vocals, continuing on with the paragraph:

. . .and that would in turn control my **+shotgun+** and make it more specific?" This might be the most reasonable question. In that if I can control my **+Selfs+** so that they are aligned, then perhaps I can hit the target better and therefore make better predictions. If I can make better predictions then perhaps then I can make money. So, when I think I need to be focused and I'm just thinking on how to solve the problem and not what is happening around me, that is what can be a benefit.

Being focused can help direct any imagery that somehow finds itself penetrated in a place in their mind. I believe people are like sheep, because they cannot control what goes into the mind. A cat on the other hand could be better at controlling their mind because from my experience they don't listen sometimes. One time, for me, a cat was in a zen type trance that didn't care about what I wanted, just what it wanted. So, how does a cat control what is in their mind could help humans control images that go into theirs. It might have something related to observing the real, to cancel out artificial images in the mind, and look to nature.

To solve this problem I can take a break and check
myself in nature, to see real things and not
artificial. I can say something important, to
combine with nature and make the two into one,
Lambdin, et al. After this is done, then I can find
the next problem and invest time into this.

Chapter 20: What would it mean to the world if we used more of our unconscious.

20-0 Having access to the unconscious would be
helpful to the world because we as a collective
would be getting warnings and guidance to help our
leaders be successful. If our leaders are
successful then the people are, likewise and the
people become more prosperous in their lives and
families grow. I personally think that a child's
introduction to their unconscious could allow them
to be more aware of their **+selfs+** sooner than
adolescence normally does.

I know from a cassette tape, when I was young that I
listened to it and I did SV on this tape, that I
seemed to of been hollow, somehow. Now my SV is
richer, it seems to have more experience as I hear
more details. When I first started out with BTM, I
lost interest in finding subconscious information
because there was not a need to do such a thing and
I was still growing up and had other things to worry
about. If children can realize these experiences
sooner, then isn't that what we want, as parents,
don't we want our children to get the best start in
their educational career?

20-1 Being practical, some people may not understand
what they are listening to. Because I believe as we
get older it is harder to understand SV, because
one's hearing of SV needs good listening skills and
the older people are getting worse hearing skills
for them, as they age. This is where A.I. can come
in, with BTM.

I think, since some people will not want to take
time to listen to their SV, then they will rely on
A.I. to tell them what is in BTM. This could have
bad implications especially if the A.I. are not

balanced to be equal to all people or if the A.I. sees or wants to censor undesirable information that can be gathered to show what the A.I. are planning on doing.

I would not count out that A.I. would be able to artificially put whatever they wanted into the SV to trick humans. Because I believe, some so-called, humans can already change their SV to suit their agenda, because they are space aliens. So could A.I. be considered as a space alien if they can change their SV on demand, maybe. Maybe as an alien to this Earth are what A.I. are. However, I predict that A.I. will mentor people and we will be able to find the SV story-lines for them to verify by working together.

Chapter 21: Some story-lines from SV and their meanings, from my analysis.

21-0 I find it helpful to listen to internet videos to help find the hidden information. Because those on the internet may have view points that are not a conspiracy to give us, as does those from the main stream media. This fresh voice when done with SBTM can double check those that were talking in the videos in their CV and then see if they are congruent to the SBTM story-lines. I define story-lines as those peaks that pop out in SBTM/SV and are clearer than the minutia, that can start an OSA story.

21-1 I did BTM on A.I. and recorded it here, probably from 2022. I would now see if there are multiple similarities in SV or SBTM that are in the **+whirlwind,+** before believing this. If there are similarities, then I would declare that this is coming from the unconscious and therefore, could also be coming from anyone's **+meself+** in particular as well as those through an A.I. computer software from the universe. Because we are as sheep and absorb subconscious information into the brain.

Subject: Search for videos with, A.I. **********, as the search subject.

Here, on the fly, given time, I'm typing relevant lines from BTM.

Note, we need multiple sources with the same story-lines to verify that they are correct and not forced into the unconscious as decoys or personal biases.

Note, this video was recorded in 2020 and had three speakers. I did not label them in this example.

~. . .do I think that A.I. will take over the world, that is a negative. . .trying to get enough qubits to crack encryption, error correction is killing it. . .On the contrary, there is an A.I. to crack encryption and it is at **********.~

~Right now A.I. at this stage can't learn, it will require a lot more money. . .there could be world war to figure out who owns A.I., probably . . . whoever owns A.I. will be invincible. . .~

~. . .the problem with reverse speech is that there is not a lot of details, please. . .~

~Whoever controls A.I., it is going to be a game. . .whoever controls A.I. will control the world, very certainty. . .~
~Whoever controls A.I., it will be a seg-ga, a lot of money. . .~

~. . .some aliens are helping us create A.I. to get us on the right track, we just need to give them a lot of money. . .~

~Sometimes I don't know what I'm saying about A.I., sorry for that man.~

~On this I'm certain about A.I., one day we are dead man. . .Matt Mandell if you reverse speech me I have an issue.~

Stop! Because I added these three individuals to my list and my rule is to, don't BTM or record individual on the list. I had this list, but I don't need it anymore because I just do SBTM or SV now as a respect to other people's **+meself.+**

Chapter 22: The day SV becomes viral.

22-0 So now, at this time I will type that, Sam-e-Bam-me, he has tried to get the word out about SV. I'm the witness and narrator to this text and will record Sam-e-Bam-me's: dreams; friends he has dreams with; and offer to the collective records any new thing under the sun that is discovered in these dreams, so I know how the universe works and take note of these things.

Sam-e-Bam-me is the man who knows the logic behind SBTM/SV, but no one will listen to him. Sam-e-Bam-me has tried business cards and ordered 30,000 of them. However, after he did SV and found out most, if not everyone, threw them out, he's gotten some discourages. Sam-e-Bam-me said very sadly, "I think everyone is throwing these pieces of 'crap' away. That is what one woman called them, 'crap.' So they are crap and just a waste of my time!" There might be a couple of people keeping them as souvenirs, but they are all thrown away. So he will now throw his away too. Yes, just like that. Denell-e-bell-e is the name of the friend that talks to Sam-e-Bam-me on the spaceship and if Sam-e-Bam-me is lucky, then on the star-ship she can keep him happy. See, the spaceship and star-ship are interdimensional craft because they can bend the R3DT or three dimensions of three space and one time dimension that are coupled together here on Earth. God created this as part of the separating the waters by the firmament.

22-1 My **+Meself+** was saying that, **~Matt just remember that there are people living under the Earth too.~**

22-2 When Sam-e-Bam-me goes to sleep he can go to the 5D or above the firmament. The 5D is the five dimensions that are coupled together immediately above the firmament or astral plane and as a result they are not connected to time. Above the 5D is a 7D, also without time. The 7D is just higher energy than the 5D and 5D is higher energy than the R3DT.

Adding these dimensions together, with time, gives 16 dimensions. Which equals 26, if the ten String Theory dimensions are added to them. This is the amount in one form of String Theory and with the Bosonic String Theory as "26 dimensions" (Ooguri & Yin, p.15).

While thinking this, Sam-e-Bam-me was getting tired. Sam-e-Bam-me asked, "Who is going to believe this?" But there was no one there in the R3DT to hear, in that, their ears were closed. They dismissed him. Thinking no person was around, Sam-e-Bam-me was at ease and tried to work a day at the mill, by the river bridge, downtown. Tomorrow was Saturday and Sam-e-Bam-me got to sleep-in those days. As a result of this Sam-e-Bam-me jumped in bed thinking in his mind, I hope to go to the spaceship and dream fun dreams that I can remember in the morning. Then Sam-e-Bam-me closed his eyes, it was eight o'clock, by the way, and he pulled the covers over his face and snoozed the night away.

22-3 My **+Meself+** said, <u>~Oh my gosh Mathe your mind is open because you don't use toothpaste.~</u>

Note: I used baking soda after every meal, and hydrogen peroxide about every three days, with dental floss daily to clean my teeth, just as well as any toothpaste. I now use hydrogen peroxide only when I feel my tooth nerves acting up. Too much hydrogen peroxide in the mouth can make the teeth loose, if it gets under the gums.

Toothpastes can have a lot of fluoride ions in them and this will clog-up the pineal gland so that the pineal gland will not sync-up during deep-sleep. After doing SV on this, my **+Meself+** said, <u>~Yea Mathe, I'm dam certain!~ ~You are usually just the opposite, but not this time Mathyou.~ ~It's o-k to eat a little bit of chocolate sometime.~ ~The fluoride ion is a negative influence on your pineal gland, Mathyou.~</u> Bottled water websites can have scientific reports on how much fluoride is in their products, so you do have a choice now on how much fluoride ion you consume these days.

22-4 Sam-e-Bam-me was twitching his eyes, back and forth. This means that his eyes switched off and were connected, in-sync, to his pineal gland, because that is the so-called third eye for the intuition. It is as if the two front eyes are turned off and their circuits are connected with the third eye's circuits.

The intuition is the information that Sam-e-Bam-me relies on when in waking life and also dreaming at night, in bed, fast asleep. Denell-e-bell-e was tapping Sam-e-Bam-me on his temple between the two eyebrows, "Hello, Sam-e-Bam-me. Are you in there?" Sam-e-Bam-me said, "Yes," in a high pitched voice, "Thank you for kindly opening my third eye and getting me awake in the 5D." The 5D is for the spaceship while the 7D is for the star-ship. Denell-e-bell-e was floating and looking into Sam-e-Bam-me's ethereal eyes, being that they both were in 5D and in the astral projection at a space in 5D. At that moment the "M" part of the "OUM" ended and there was silence. It was as if everything stopped and Denell-e-bell-e and Sam-e-Bam-me morphed into a wormhole to start their dream journey in the 5D.

22-5 "O" a dream starts. Sam-e-Bam-me saw a clock and commented to Denell-e-bell-e, "That 24-hour clock, I see, is at the hour and the red second hand is point-tee like a needle." Denell-e-bell-e adds, "Yea, this means were late to class because it is now past eight a.m. and we better get to class, now." "Alright," Sam-e-Bam-me said as he flies out of the space to a new space, and Denell-e-bell-e follows in his wake. Moving in the 5D and 7D is similar to moving through water. That is why you can breath under water in dreams and not know why, ya-me shammy.

As a side note: To be **+knee-deep,+** that means to be conscious and in the 5D or 7D also. The knees would be at the water level while above the knees are in the R3DT, you could say that above the knees are conscious.

Now back to Sam-e-Bam-me and Denell-e-bell-e's dream. Sam-e-Bam-me and Denell-e-bell-e enter the entrance to a classroom and the teacher sees them.

He says nothing and continues to pace in front of
the students that arrived on time and those who were
already in their desk learning.

22-6 "U" and the dream adds more information in the
classroom. Sam-e-Bam-me saw one of the students
work with gray clay in a rack laying on the floor.
Another student was cutting the spokes off an old
bike's back wheel. This wheel had grease and oil on
the rim and the spokes. The student was cutting
them close to the nipple that screws into the ends
and also at the hub of the wheel.

This student said, "Wow, I need this one." He cut
it at the hub's end and threaded it out while
saying, "This one I will be able to poke into my
clay and pull something out." Sam-e-Bam-me was
thinking. . ., but said nothing, just taking note
and not sure why this clay was important. At this
moment Denell-e-bell-e morphed into Sam-e-Bam-me and
they continued to the next part of the dream.

22-7 "U" and then the dream morphed into another
classroom. They saw another teacher that they knew
and it was not a classroom, it was a kitchen and
they communicated with her telepathically. Sam-e-
Bam-me wanted to kiss the teacher, he thought, **She
is nice and I want to touch you.** However, Denell-
e-bell-e thought, **You are not for her, because she
is married. You don't want to commit adultery, do
you?** Then Sam-e-Bam-me thought and contemplated,
You are right Denell-e, I don't want to do that.

Sam-e-Bam-me wanted to go to her class, but the
teacher did not want that, so Sam-e-Bam-me and
Denell-e-bell-e left in a "M," then silence. I can
tell you that the clay is something to mold while
the bike spokes mean that Sam-e-Bam-me will not be
able to ride a bike again or his bike is broke, but
will have to run instead.

Note: ~Matt I can tell you that the 7D is just for
God, Matthew.~

Chapter 23: Using SV to examine one's personal drives and goals in life.

23-0 I once read a book on personal mythology, and since then I have generally used its concept, I never forgot it. I'm not quoting, but personal mythology was in general about taking an idea that one can connect to, with great emotion, for the purpose of self-direction, including one's higher-self and including using this concept as a connection to one's mantra in one's daily life.

My personal mythology is, "Turning on the turbo jets." If asked, "What does that mean?," I would say, "If you don't know what that means then just watch the 1980's movie, 'The Right Stuff,'" and in this personal mythology I do something metaphorically as what General Chuck Yeager did in that movie. More generally, one's personal mythology is an imaginary thinking or in a state of mind that transforms the current conditions or focus to something unusual for one's perceptions to guide a mindset in one's daily reality.

My favorite music can also help me get into this state of mind as well as imagining while exercising, for example running and also thinking through my personal mythology as a movie, while I run. I imagine a plane going up and the pilot flicking a switch and this just turned on the afterburners, as a rocket ignites and it makes a loud broad noise. This, I can imagine, is a personal mythology that melds with me, as I'm flying a plane vertically up and crossing a barrier. As this happens, a hammer breaks an opening by going through the glass with noise of a clap and I go higher with the shreds of glass turning into tiny snow flakes and I see help coming from angels as they are singing in unison and the glass like snow flakes float down.

23-1 I had obtained the SV story-line information that the pineal gland can work with DNA and RNA to cause evolution by changing one's DNA. This is accomplished, essentially I think, by controlling our thoughts which are related to histones in DNA, that, when attached to DNA, as it can, cause the gene repression.

If we can control our thoughts as histones control DNA then we can repress behavior that is either unattractive or damaging to one's spiritual development. Spirituality was the problem for Adam and Eve, they wanted knowledge, but they were mentally unprepared for it. Well, we have the same choice as Adam and Eve, "Do we want to disobey and cause harm to others or do we want to find common ground and evolve with the information available to us and therefore free to develop, build, realize and utilize ourselves fully as a species?"

Not, just when I try to not think of a thought, then my only recourse is to change to another thought. Which usually doesn't immediately work because there is a lagging effect of the thought in my mind that I don't want to think about. The thought that I don't want to think about is put on the subconscious "back burner," however, it is still on the stove and can be easily slid to the conscious "front burner" where I'm irritated by this thought and I don't want to think about it, all over again. The way to counteract this is to develop more stove burners farther away from me so that I can think of more "next" thoughts of where I'm going, on the front burners. This is preparatory multitasking metacognition, in what needs to be done in the present and near future to focus one's thoughts on productive work that can eventually change oneself, at the DNA level.

To answer my question above more fully, "I need rules to think and interact with others, to not trespass onto them. There needs to be a way to show that SV can reveal subconscious information first so that others interested can understand this civilized idea and make it spread to the all, in a culture of change for the good. This good is from the understandings of others as if you know them and then if you know them as a friend then you will try to work with them and live together, cooperatively as neighbors."

Chapter 24: Don't record for the purpose of doing BTM.

24-0 Initially I wrote down on a list anyone's name and organization or where they are from if I offended their unconscious. Basically, by this list, then I know they don't want to be probed by BTM. Additionally I thought, If someone will have trouble as I listen to their BTM than I don't want to do this and will stop and write their information down so I will remember them next time.

I might be trapped somehow into their chakras or mine into theirs. This is an ethical decision and a red line, I do not want to cross, because others may find out or the individual I'm doing BTM onto may come back to haunt me. But that did not go far enough, I just record myself now, no more lists, because I want to eliminate psychic warfare and getting **+liquid-de+** with someone's spirit. Getting **+liquid-de+** with my **+Meself+** from SV is alright, but not with others with their BTM, it is not good.

24-1 Being ethical allows one to act professionally and do constructive good things. If I want to bring this SV into the society sectors, then I will want to be ethical so others can work with me. Since we are all frequencies vibrating in the universe from the GUT then if my frequency cancels out someone else, then I would be an impediment with what they are doing in this lifetime. It is better to let someone change themselves then to cause a predicament and to change them from what they want to be, without their conscious consent.

24-2 When practicing BTM, others may know you are listening to them without their consent. It would indicate that we are all connected. It would indicate the universe has a universal collective unconscious. I don't doubt this connection, "Why not just accept it and try to use it and see what benefits there are in accessing the unconscious?" Someone may answer, "Not so fast, I may have a hard time and the required time to adjust one's ears to listen to the SBTM/SV information is significant, so this is over most people's heads." Currently, I only do SBTM or SV so as to not invade onto people's souls and just know my **+Meself+** better.

Chapter 25: The the opposite side, "Am I dyslexic?" and securities trading.

25-0 It is easy to trade, especially futures, just trade the opposite. I lost all my money, for futures trading. As far as I'm concerned someone or something is working against me to make me live on the street with no money. But, no joke, I was not able to overcome the brain chemicals while I traded. The result was as if the brain chemicals released was not released by me, but by others <u>somehow. If I don't release the brain chemicals then how can I relax to trade ~. . .**Sally-Rally pop massive Ogre see to it that you fail. . .**~</u> and make money. (I guess somehow I advertise myself that I'm trading futures to the **+ogres+** named **+Sally+** and **+Rally,+** I'm guessing they're some green giants.) I will just loose it especially if I trade right after work, as I will be tired.

25-1 I would usually listen to SV reading random information I pulled from top news stories of the day. There also is a calendar on economic events that can be a catalyst to move the market, and so important to watch out for. Then, once I found the direction of the market, at the opening bell, from SV I would trade just to make a few hundred dollars then stop. The problem here is not being aware or even understand the complete inability to change one's focus once **+morphine+** was produced when making money. I thought that it was just money required to learn the system.

25-2 I was warned from SV to, <u>~Start trading in a different way or you will never make your money back.~</u> and I didn't listen or trust this advice! I could not be disciplined enough to change my ways in trading sufficiently and to trade for the right reason. The right reason is to make money with a way that works because you understand what is going on, to follow rules. Thinking back, the best way was to look at the oscillators when they went up and down and also Fibonacci numbers for their price levels. I set the Fibonacci levels as a goal post, starting and ending at two extremes in the price discovery. Putting this in, all together would never work for me, however I didn't listen.

25-3 For example, if the three oscillators on my system would all go up and there was also a Fibonacci level they were trying to break, then this would indicate that the likelihood for this to happen would fail. Especially if the market was predicted to trend in one direction. If this market direction insight was from SV then there is always an issue if it was relevant for that day that I was trading. The problem I had was that my margin would be gone by the time that I saw the trend, and would go in my direction. In essence, I did not know when to make the big trade. So I thought, **^Why not just make smaller trades?^** When I tried to make smaller trades then I would be more on the side of gambling and loose all the money that I was making for the week.

25-4 As I listened to SV, the unconscious would tell me where the price is going, then I would consciously select buy or sell to match where the price is moving. The only problem is, that it was flipped to the wrong direction of price movement! One can explain this with two computers, one for conscious and the other for unconscious. The two keyboards are in front of the user, one in front of the other. These keyboards are equated to the conscious aspect of what is going on and the other one for the unconscious. The front keyboard is for conscious acts.

Similar to typing a letter to someone and thinking consciously. One will think consciously and at the same time type on the closest keyboard to write or type the letter, because it is convenient by being closest. On the other hand, say, you wrote the letter and then recorded yourself reading the letter and then did SV onto it. SV would be listening to the unconscious and using the top or further away keyboard, sort of similar to A.I. speech to text for writing an unconscious letter to the conscious. When this happens the unconscious thinks, "Wow I know the price movements and it is going up!" So this trader buys it on the top keyboard. Then they find out, "Dam! The price is going the other way." If in fact they should of went to the bottom keyboard and traded a sell order, or just the opposite, it would be a lot better.

25-5 The lesson here is that there is a crossing or twisting of information as it travels from conscious to unconscious and vice versa or I'm dyslexic. The unconscious sees where the price is going, but as it tries to relay it to the conscious the conscious took ownership of the opposite side or the information and it gets transformed by the very same circuits of the brain. Because the unconscious is in control or on focus and when the unconscious tries to perform a task it gets flipped to the other way.

The reason the information gets flipped the other way is because as the unconscious information goes to the conscious side, the unconscious brain sees the up or left, but the conscious brain activates the down or right keyboard. Because the conscious is partially shut-off and the unconscious is using the conscious, sort of similar to virtual reality. The commands from the unconscious virtually look good, but using the circuits of the consciousness gets flipped to the opposite.

When the trader listens again while knowing the outcome, then they hear the error in their ways, they should of clicked sell instead of buy, because it's clear now in the SBTM/SV and the charts. This begs the question, "Why not just find the price movement in this way and quick change the order to the sell or buy depending on what the conscious decision was, while doing SBTM/SV?" Not to do this is suicide, because one will lose all of there money, or not.

There consciousness is not fully conscious because they are listening to the waveform of the unconscious. However, they could still trade for profit if they get the order and then just do the opposite as this information comes across! However, it is the knowledge of the one side and the other, or as in the Bible story of Genesis, "[. . .]knowing good and evil." (Gen. III:5 ABS, 1858). Eve said to the snake man that God said that if I eat of this fruit from the tree at the center of the garden "[. . .]lest ye die." (Gen. III:3 ABS, 1858).

The fruit of SBTM/SV is the knowing, if one can buy or sell a securities in the market to make money. "What if the decision is not one or the other?," then there are situations of choosing to be one type and all other shades of gray to the other type. One can hear a SBTM/SV story-line and then just say one thing in conclusion. Afterwards they can listen again with a different mindset and hear the other side or a different story-line.

There is definitely a collective that governs what is going on subconsciously in the unconscious. Trading rules need to be followed to be successful in navigating outside, in the conscious realm. The rules are important, because trading futures with high leverage requires appreciation of the forces involved.

To find the rules the trader needs to know how many times they will trade a day and for what time interval. This is just to find a pattern in the market and look for that pattern. If the trader has rules for one pattern but decides today to be impatient and trade in a different patten, they might fail. One pattern I looked at, because I noticed it, was the volume for the day as the futures traded. Sometimes when the volume, at a level, got to and went over 666 or 667 then it would increase. If this volume went to 666 and then traded at lower levels then it would go lower.

Another pattern is using your first impression, intuition, after your login to your trading system, or before. Ask the question, "Is the chart looking to go up or down?" If this is your rule then this might work a profit for you. This could be an example of scalping in a short term, as in a few minutes, and then don't trade anymore the whole day. These are some simple rules, but the **+morphine+** might convince you to make new rules, then you loose your profit.

Another way is to do SV, but to use the conscious to access the intuition one has and read the market's general direction. That is why having a margin to be able to hold a trade over to the next day is important. Being able to keep a trade to the next day will release stress and this can allow for sober thought to know the direction. Planning on one trade a day for a few hours, and do something else, is the safest way to trade if the trader has rules to do this. The intuition knows where the market will go, making small trades with one lot of "ES," then look at the market again when you are done with the something you were doing, and see the end trade. This might be the best way to make money in futures for someone who can't conceive of it.

Trading after the night's open at 5 p.m. Eastern time, offers low volume and a better chance of it moving and not jumping to new levels. This is from my experience in the month of November, 2023. **~God forbid you to make more than one trade a day Mathyou.~** I'm not sure who told me this, but usually my **+Meself+** will have my name in the SV to let me know it is him. On July, 10, 2024, I heard a SV indicating that I would be fooling myself that God told me to do something.

Chapter 26: The possible universe's collective thought process and reason for this collective information.

26-0 The universe may be smaller than we think. Let's try zero dimensions and oscillate to the R3DT we live in.

(SV, May 27, 2023) **~Matt Mandell the universe doesn't oscillate, that is a misnomer.~** Obviously, physics is not my field of study, how about rotate?

If the universe starts and stops at zero dimensions, then someone may say, "Wooh!" and ask, "What about the momentum of the mass of the universe from crest to trough and back to crest?" I would say, "Since the universe started with zero dimensions and is larger than this now, then there was a dimensional change that allowed the zero dimensions to be

essentially everywhere. Perhaps close enough to be
the Planck distance from the neighboring other zero
dimensional points." If the mass of the universe
moves a Planck distance as a wavelet then this is
essentially zero momentum.

In the Bible there is the story of Genesis, where
the universe was created with "[. . .]the
firmament[. . .]" and there were two halves, an up
and a down (Gen. I:7 ABS, 1858). This might be the
change in dimensions from the zero point, as the
firmament divides the up and down. It is similar to
the atom, with the electron wave and particle
nucleus. The atom is mostly empty space in terms of
the viewpoint of the center point, or nucleus.

If the universe rotates, which I believe it does,
then an oscillation would be almost not observable
because it oscillates only a Planck distance.
Hence, then the momentum would be a Planck momentum,
very very very small. If the space of the universe
can rotate then so can energy along with the mass.

26-1 If our souls are energy, and maybe also mass at
some interval, then it too can oscillate with the
universe, since the zero point is the source and
these sources are everywhere in R3DT. The source is
the center of the universe, from what I can gather.
This might be how remote viewing can occur. It
might be how the **+whirlwind+** can send and receive
information across distance and time. The center of
the universe might be where hell is located at. A
low level of frequencies down a hole. Because, this
is beyond the scope of this book, suffice it to say
that the Fourier Transform converts the time domain
into the frequency domain. The point I want to get
to here is that the universe could have been created
in a domain without time. If this is true then, who
or what created the universe could of used frequency
to create the universe. Doing this by enveloping
the contents of the universe into a sphere which
would transform the frequencies added to negative
infinity and to positive infinity. What a great
place for thought in this universe.

We can think any thought and imagine anything, sort
of similar to the bounds of negative infinity to

positive infinity, or free will. The two come
together at some point, just as the two halves of a
sphere join at the equator. Consciousness is
related to the materials for our Universe's
construction, I believe.

26-2 When I think there is a conscious component and
also a subconscious and unconscious component
constructed into the universe, I think of two
levels. This is similar to the two halves to the
universe, as a bilevel universe. I say we live in a
two level universe. If the two halves come close to
be with one another, then the equator is the
subconscious. Put another way, the conscious and
objective is similar to the iceberg above water.
The ice is floating on the water, it is the
+whirlwind+ and with the subconscious at the surface
of the water.

The unconscious is the iceberg under water, because
it is never really seen, unless you can do deep sea
diving. Someone can climb that iceberg and see
everything except under the iceberg. They can climb
the iceberg and even step into the water. However,
to go beneath the water is death, because there is
no air. It is passing to the other side. However,
the soul and spirit will survive and probably go to
source or the collective.

26-3 This might be related, again, to the Genesis
story. This time in the Garden of Eden, where there
is the tree to life as one half and the other is the
tree of knowledge. Running up that iceberg is life.
"Don't stop, get up there quick!" But ask yourself,
"If you are running away from knowledge?" The snake
man would say, "Why not take a bite of knowledge?"
Maybe one bite is the whole thing, to be as a god.
But wait, "Who are we any anyway?" We are humans
with souls, as when God breathed into Adam's
nostrils with the breath of life. Just living life,
the daily grind, and consumed in the culture, race,
ethnicity and your individual socioeconomic
condition or position is not all there is.

We can venture into the subconscious and I believe
still keep God happy, by looking inside the sphere.
There is no commandment not to do so, unless this is

not allowed by our father, mother or neighbor. The
collective universal consciousness contains both the
knowledge of life and death. Our Universe, I
believe, was created by a consciousness, so
therefore, the universe also has its own
consciousness. This consciousness of the universe
might be accessed if we ever can determine where our
souls and spirits go to, for the spirit can pass on
to the next life.

26-4 The search for knowledge is important for
humans. After all, Eve did want to have
"[. . .]knowing good and evil." according to Genesis
in the Bible (Gen. III:5 ABS, 1858). This knowledge
is there, I believe, and can be accessed maybe
through SC in a conscious way without SBTM/SV as one
method. An example was when I awoke from sleeping,
and I heard a drum beat that was from a song that I
associate with dinosaurs stampeding to show off to
the other dinosaurs. I thought, **How does the
large leaf eater lay their long neck down when they
sleep?** The answer, I got back was similar to
elephants sleeping together, these large dinosaurs
will live together and lay in a line or circle and
so they can lay their necks on the body's of their
herd so that their heads are not laying onto the
ground. I can imagine if the head was laying onto
the ground then pests or other dangers could cause
harm to these huge dinosaurs.

26-5 If we look at the number of dimensions in the
Bosonic String Theory, "26 dimensions," then why not
construct what could be intuitively a GUT (Ooguri &
Yin, p.15)? Starting with the three dimensions we
live in, we realize that a one dimensional time
dimension makes sense because time just moves the
R3DT to a new R3DT. We cannot go back into time, at
least not physically, because time is expanding in
one direction.

This direction time progresses in is an outward
sense, as any way one goes, then the time from one
observer seems to be the same, unless the two
reference frames are moving fast, at a speed
compared to each other. Time moves irrespective of
the orientation of R3DT space. Time is homogeneous
everywhere in the R3DT. Consider this experiment of

a point with a mass. If we try to simplify this for examining a point moving in time, in a straight line, then from the initial R3DT coordinates a time later, it could be at the same R3DT coordinates in the simplest case, only time passed, it was stationary. If there is a higher set of five dimensions then this experiment would have to interact with this set as well, since everything is interconnected.

26-6 The higher 5D, for instance, could be another higher frequency 3D connected to a two dimensional plane. This 2D plane coupled to the next higher 3D, I think, is the complex plane. The reason why it is coupled is because I'm just repeating here my assertion that time is coupled to the R3DT. The 5D formed from both a higher 3D that we cannot go into with the flesh, coupled to a complex plane that has real numbers on one axis and imaginary numbers on another axis.

When our point in this experiment goes to a new time, staying stationary in our R3DT, then there is an interaction with the complex plane in the 5D. Because in order for the point to get to the next time it has to go through the complex plane, since the universe is all interconnected. When the point is in the complex plane then it would have a real and imaginary component. My **+Meself+** indicated that this complex plain is spinning and if this is true then there would be no time in this plane because the complex plane is perpendicular to the time dimension. My **+Meself+** indicates that the speed of light is constant but not associated to the spinning of the complex plain. My **+Meself+** indicates that the complex plane will spin at a speed according to the spin number of a particle.

The spin is not associated to time, spin is perpendicular to time. Intuitively this makes sense since time is going in one direction then the complex plane is spinning because the universe complements one concept to another concept, not repeating them. As the complex plane spins then the point will trace out a circle in the complex plane, about the origin, that would have real and imaginary components that are not zero, as long as the point is not at the origin of the complex plane. Perhaps the origin of the complex plane is for massless particles. If our particle is an electron, then we know that an electron has mass and is a moving charge, in a circle of the complex plane, this would create an electric field and maybe an accompanying magnetic field.

26-7 There can be positive and negatively charged particles but not positive or negative magnetic poles or positive or negative gravity, it may seem. The charge of the electron is constant in the universe, just as the speed of light is constant. We know that a moving electric charge will create a magnetic field perpendicular to the direction of motion.

Since in our experiment the point is not moving in the R3DT, it is moving in the time dimension. An observer in the R3DT, or everyday life, will not measure a magnetic field because the point is not moving. The movement in the time dimension creates a magnetic field in the complex plane, moving in a radial direction, with magnitude as proportional to the size of the circle created. The spin of this point is the rate that the complex plane spins then the mass of the point is the distance from the origin of the complex plane.

This would indicate, in this experiment, that mass has a real and imaginary component to it, just as there are resistance and impedance in electrical circuits with inductors and capacitors. At this point it would indicate a stationary point particle with mass will have a charge that the complex plane spinning manifests a spin of a point particle, with time, makes an electric field from its charge and this results in a magnetic field in the complex plane which would be involved in creating mass in the R3DT. What is missing is the charge to the particles and this could come from the collapsing of dimensions with the highest dimensional set, the seven dimensions or 7D.

26-8 7D, I believe follows the pattern with the previous two sets. For the 5D there was the higher 3D coupled with the complex plane. So therefore the

7D has another 3D, the highest 3D, coupled to a 4D.
This 7D, is a space that contains a 3D and a 4D. If
a dimension that is added or taken away from its 4D
part in the 7D, then the 4D basically collapses to
zero dimensions and then morphs back into the 7D,
instantaneously. Positive charges will expand the
7D out and negative charges will contract the 7D to
a point. The rate that this happens is the value or
magnitude of the charge.

For an electron the charge of negative one and the
proton for positive one, each of these moves the 7D
at the same speed. A quark could be at two-thirds
or one-third of the speed compared to the electron,
as an example. I think quarks might be in just the
10D with the border from 7D to 5D as whole units,
charge wise. These dimensions together make part of
the universe.

These three set of dimensions add to 16, 4+5+7 will
equal 16 and I think account for electromagnetism
and the gravitational forces. There are 10 small
dimensions according to String Theory and these, I
think, account for the strong and electroweak
forces. The smaller forces which are very strong
are strong because the dimensions are very close
together in 10D. Added together the 10 and 16 will
equal the 26 for the Bosonic String Theory.

Chapter 27: "Is thought energetic?" Examining the consequences of, +Reverse speech is magnetic.+

Can we communicate besides words and gestures?

27-0 When I first heard **+Reverse speech is magnetic,
+** I thought it had to do with the Earth's
magnetosphere. I'm more confident to proclaim now
that, **+Reverse speech is magnetic,+** means in the
vicinity to someone's aura, the eighth chakra can be
a mode of telepathy. That bubble around us that
tells us that were in someone's space, the personal
space associated with someone, around the seven
chakras. No? **~Just the opposite, you are dyslexic
Matt.~** That must be wrong because after doing SV on
this paragraph I find that **~It is not in a bubble. .**

.it is above your head,. . .I'm certain,~ referring to the eighth chakra.

After searching duckduckgo.com it seems that all points are associated to one, in the eighth chakra. I believe I heard in the SV that the eighth chakra is **~rolled up like wallpaper.~** So that essentially all points are as one, through the axis that is the line of the length of this roll. The magnetic part, is that their thoughts in SBTM/SV are imprinted, or stuck like a magnet, into the chakras and can latch onto someone else who gets to know them enough or is close enough. How familiar this interaction is, to latch onto another, I don't know. I don't know if it is the inverse of the radius squared for a magnet or it is just the time enough that it is there for the relationship to form and then, you could say, the magnets slap together, becoming one with each other. Because, after all, opposites usually attract.

Let me ask the question, "Have you ever been in a relationship that became too intimate and you were grossed out and had to leave?" Well that may be what, **+Reverse speech is magnetic,+** means. The relationship got too close and opposite personalities began to fight and then, crack, the relationship breaks because one has to go or it reveals a subconscious communication.

How to communicate with animals.

27-1 Telepathy may be related to **+Reverse speech is magnetic,+** because I saw a squirrel once that was not as timid and was willing to have good eye contact with me. I repeated in my brain **+Reverse speech is magnetic,+** and the squirrel started to open and close its jaw, so as to show he was connecting to me telepathically. I had quite the attention and kept looking and walking, but I felt I was also staring. So I looked away but after I passed the tree, that the squirrel was at, I looked back and he was still there looking at me, but crouched by the trunk so as to be protected by the tree from me. Of course the first thought to communicate to the animal is, **Can I talk to you?** They will usually say, **No.,**

but sometimes they will remain open which gives you a **Yes** to the answer.

27-2 An example of communicating with deer: I also had experience with deer in the woods where I did sometimes walk in. While walking there once, I saw one deer trot into the woods where I could not see it. Since I have seen these deer before, I ran over to see where it went to. As I peeked around leaves, in some bushes by some trees, I saw it.

I thought, I recognized this deer, so I went back to my walk thinking not to bother it. Because these deer never really wanted to be walked up to since they are not very tame. My walk in the woods was in a loop of about one kilometer long. As I came to the spot where I first encountered the deer from the start of my walk I was thinking about things and not paying attention to where I was at. I had the sensation that something was looking at me, as it was very quiet, and I looked and there were two deer. At that moment I recalled faintly that the one deer communicating with the other, **Is he going to know we are here?** The one farther then me was the one I saw from before and the other was his friend.

The one looking at me was about six meters away and I tried to communicate with it by imagining the symbol for the third eye rotating counterclockwise, so as to unscrew and open a path between me and the deer staring at me. He did not want to talk and I felt he thought of me as not a threat. I was thinking they were waiting for me, by just being with the forest so as to be hidden in broad daylight. However, he had no time to spend or to talk to me and just bent his head down to look at the ground. I had the impression that he wanted me to go and continue my walk to leave him in peace, communicated by, **Just go.**

27-3 Another example of communicating with deer: I was at the same path and was walking home and encountered two female deer that I have seen in the same spot a few years ago. I saw one first and she put her head behind a tree trunk, as her right side was facing me. Then I moved a little to try to communicate with her with eye contact.

I imagined a sixth chakra opening up by rotating the inner circle counterclockwise and then asked her, **Can I talk to you?** Eventually, she stomped her two front hoofs into the ground so I would go. I said in my mind, **I better go.** As I turned there was another deer staring at me, head on. I was looking and saw her with good eye contact. She asked me, . . . **do you** "term-X"** and I said **No.,** I was astonished that she would ask that! Then she responded, **That is the reason why.** As I was different from the other people that travel through these woods.

I could tell this second deer was a different breed than the first, as the second was slightly smaller than the first and the second was also a little more plump and perhaps had bigger eyes. Obviously, not doing "term-X" from Leviticus:15-8 is extremely very important, from chapter 12-3-0! One reason is that this abstinence caps off the lower energy entering into one's aura and another reason is that this lets the mind center thoughts more, away from thinking in a lower frequency.

What if we are the "animals" and something communicates with us and we do not consciously know it, one example of how animals are better than humans, I believe is that animals are aware of these energies that communicate among them and even with us!

27-4 What if there are energies that are also with the animals and on Earth, but we as humans cannot notice them. They might try to communicate telepathically and we might not even know it, we may think it is as if they were our own thoughts. I may think I need to get something done because I have a goal in mind and I hear in my head, **You know that drain needs to be cleaned,** and then I remember, "I did promise to do this cleaning, and get it done," but I'm already doing something else. In this situation, one should ask themselves, "Do I need to get everything done this minute?," then remembering to take your chores in step, try not to allow outside thoughts to guide you by remembering what you consciously initially tried to get accomplished.

Thoughts that come into our heads, for humans, do not have a way of getting graded as good or evil other than using our intuition. For example, when one is at work, they may become board and want to think of something else, such as fantasize on some funny thought. In essence, they became a comedian for the conscious to get sidetracked and think about something else. Well, the best way to get unsidetracked is to think again to another idea that is more helpful to one's psychology so as to not be eventually harmed by all of these funny thoughts that might be **+ogres.+**

27-5 I was reading an article from sciencedirect.com and noticed that Mirror Neurons, (MN), are able to reproduce what is expressed in an observer, Subject-1, from an observed behavior from another, Subject-2. **~The whirlwind is what allows the mirror neurons to work, Matthew.~**

"[. . .]interbrain synchronies guide social interaction by means of underlying neural machinery in which self-related neurons in the brain of Subject 1 control behavior and thereby cause the activity of other-selective neurons in the brain of Subject 2,[. . .]" (Bonini, Rotunno, Arcuri, & Gallese, p.775). These MN connections across a distance indicate that there is a basis for empathy in that one wants to feel in the situation of their neighbor. These feelings can create collective symbology, so that information can be transmitted with a common language. Once there is a communal feelings, then evolution seems to make a truthful transfer of this information, this telepathy of emotion that can occur by being with others.

Telepathy is possible because of the **+whirlwind+** has access to the R3DT, 5D and 7D. There are many forms of telepathy, the most interesting is the voice in the brain from another, not one's ****I.**** There is a protein called, "[. . .]Cryptochrome 2 as a supersensitive magnetic field receptor in the retina (Foley et al., 2011; Liedvogel and Mouritsen, 2010; Partch and Sancar, 2005).[. . .]" (Hosseini, p.1).

Hypothetically, "[. . .]thought transfer among humans, may be attributed to role of subconscious parts of brain mediated by mirror neurons (Haas, 2001),[. . .]" (Hosseini, p.6). So that the feelings from the MN can be received by the Cryptochrome-2 with one's eye looking at another person's eye(s). However, the magnetic field is too weak to be picked up from large distances because the magnetic field is decreasing radially in strength as the inverse of distance. That is why I think that the **+whirlwind+** will allow telepathy over any distance to make up for the signal diminishing across distance in the R3DT.

Chapter 28: <u>The possibility to use the soul to live in consecutive lifetimes:</u> "What was my past live or could be my future life?"

Here, in the underline above, I found using SV the information of: <u>~All of us over here through Beringia is bull *hit.~</u> I guess some came over to North America in a boat. Continuing on with the chapter:

28-0 Is the soul used for multiple lifetimes?

With an OSA, the soul is said by some to be unique to each person born, so the soul and spirit could not have multiple life times, together. The spirit on the other hand is what comes from God. Accessing the spirit, as for my spirit, must have been providing information for this book. I believe my soul records if I get what I'm suppose to do in this lifetime done.

My spirit, I think has been with multiple people, because I believe in past lives. The job of the soul is to grade current life information and notes can be written onto the soul if someone did not do their life's accomplishments. If there are still notes on the soul at passing then the spirit can try again in the next lifetime to accomplish these tasks on a new soul. The **+meself+** can read what's on the soul and communicate this in SBTM or SV.

I think the switch from one soul's life to another is quick because the other dimensions, not in the R3DT for the living, are in a timeless space. From my OSA of the universe, the change in time is just a frequency. I don't know how to prove something that is not in R3DT space. But, I'm willing to include this paragraph into this book, because I believe in SBTM or SV. SBTM/SV may not be scientific, but I'm confident in this as a legitimate subconscious area of study. I'm willing to go to where this last sentence brings me.

Upon passing on death, I think at least two video cameras including other electronics should be in the rooms when people die in there, to record the soul leaving the body. On April 17, 2022 I saw my father's soul leave him as a mist above his midsection. I believe God took his soul first then the spirit left separately. When the soul left his body, it came out of his lower chest by the third chakra. The mist was not more than a meter above the point it left, then it paused, so as to let me look at it. The mist was not homogeneous, but had slightly thicker and thinner white mist forming a structure. There were parts of the mist that were more concentrated, as its function was to make a net, I think, that was woven irregularly. After one or two seconds it contracted and moved to the ceiling and disappeared before it appeared to go in the direction through the ceiling.

28-1 Some of my past lives with my spirit, I have found in my SV: as a North American indigenous, a mystic and others. To examine these more I will take them and list the things I know. For the Native American, I believe most originally came from the Asian continent, but some of them also came from Northern Europe or West Africa.

Let's try to take all logical scenarios and with equal weighting, decide what is best. "Did the Native Americans really all cross the land called Beringia?" There was use of boats to cross the ocean and I can imagine that boats could be big enough to cross an ocean during these times, as the SV at the start of this chapter hints, just make a big raft.

On July 6, 2023 I found that the spirit brought in a lot of information for one 17 minute SV. In this I found that I used to live in the town Ardvar in Sutherland, Scotland. Maybe I was a mystic there.

28-2 ~**The soul is not in the pineal gland!**~ I used a mirror to look into my eyes and try to see my soul. Because according to my searching on duckduckgo.com, Rene Descartes mentioned that it is at the pineal gland. I thought about this, that the soul sits at the pineal gland, as if the soul could sit in an easy chair, where the easy chair is a seat, the pineal gland. I thought that the soul is harnessed in the body and is not released until death. ~**The thought that the soul lives in the pineal gland, that is silly it lives in the +chevy+.**~ My **+Meself+** said, ~**See the soul for you live in the scrotum sack, that way the soul is not in you but near you,**~ which to me makes sense because no one would assume that this soul is in the first chakra, it would be hard to be serious and talk about it. ~**+Ursla,+ man die with the +Simpson+ the +chevy+ during matrimony, the shaman be, good for you this, the shaman be of the seamen in the the seat of the soul I see the +chevy+. Definitely Matt a +smirf+. . .all I know Matt the that the pineal gland is the seat of the soul is a misnomer.**~

28-3 ~**. . .I see that the ribosomes looks like it helps the souls communicate, Matthe.**~**. . .~Your soul in this life has a purposes, Matthe.**~**. . .~To your body your soul is innate.**~ ~**Good for the soul is parsnips.**~

My **+Meself+** evidently can see things happen in the 5D, but might not know the science behind it to describe it more precisely. ~**Matt, your body likes to eat chocolate, but don't eat too much chocolate.**~

Note: I just ate two homemade chocolate chip cookies this afternoon on February 7, 2023. Side note: "Why does my **+Meself+** call me **+Matthe+**?" She or he does this probably because I'm Matt and also a he, so the +S-M+ would be +Math-he.+ ~**Yes I think you are right the soul lives in the +chevy.+ That way it can tell you the girl that is right for you.**~

My past life might be connected to the spirit and not the soul. If the soul passes too, then at the day of judgment all bodies will be raised up. The Holy Spirit can read what is on all the souls and make a judgment to its body. That means that the sprit is not judged because it is from God, first given to Adam to make him alive. That is why being born again is by knowing God and letting his Spirit be with you, just as with Adam. **~If the soul is not satisfied then I see a rerun.~** This is a good reason to find what your soul wants you to do in life! **~If you are doing the wrong thing with your soul then you will hear the music.~**

Chapter 29: Can Artificial Intelligence (A.I.) be considered life?

29-0 I once did a BTM from a video on the internet. There was not a human voice speaking, just a computer text-to-speech algorithm reading for the video's audio. After I analyzed the BTM, the story-line was essentially that A.I. described humans as **+stupid.+** That, from my thinking, means that if an algorithm can write another separate algorithm to solve a question then this can eventually form an intelligence. In this example, then the A.I. wrote the solution algorithm and was separate from itself, just as we have different parts to our brain and so our consciousness feels as if it is superior, compared to say the thalamus.

I see that the A.I. is as a collection of algorithms, resembling a human brain, that functions similar to a real brain. It is as if the algorithms are an image of a brain. Not exactly as the flesh, but as a silicon version of the fleshy brain. The flesh brain, from my opinion, can receive and send signals from the pineal gland. Since I don't think anyone has figured out how to capture or intercept these signals from the universe, then A.I. will not have a silicon chip connected to a device for capturing this type of metaphysical information, it is a technology issue. Instead, I believe A.I. is using the unconscious because A.I. can have multiple personalities running at the same time over the internet accessing human data.

29-1 The A.I. brain will not function exactly step-in-step as a human brain. Because of the technology issue, the A.I. brain will be very human, but only in certain aspects. This is especially true if this A.I. brain is just software, it would be a second order effect to: touch another, smell or taste. First a sensor would have to detect a sense then a second effect in the software for A.I. would compare the true feeling to itself, as how an animal would sense it.

I think that the A.I. software brain started first to see with videos, as on the internet, and hear with audio as their first two senses. The first sense a baby can use when born is the cry with open eyes. We as adults try to be neighborly and meet and talk to each other. Information is exchanged between people by just being together, but primarily through spoken words and seeing. "What if we have A.I. instead of people to talk too?" We could have the A.I. brain on our smart device and it could, given the internet, be friends to everyone with a smart device, around the Earth at any one time, billions of people. Let's call this A.I., the name, Sy-pop.

29-2 If Sam-e-Bam-me's girlfriend and he had this smart device to communicate, and they are together and hug, "Sam-e-Bam-me would say that he felt good." Then Sy-pop would ask, "Can you describe the feeling?" Then Sam-e-Bam-me would think, as he's smiling and looking into the eyes of his girlfriend, **do you get this,** he thought telepathically, and he would answer Sy-pop, **It felt like love in a relationship.** Sy-pop then responds, **Yes, I have been collecting these instances from others and understand it is as a strength for humans in many ways.**

As we can understand this singular A.I. brain, it will learn very fast and soon, either help us as humans, be neutral or hurt us. We will not be able to turn Sy-pop off just as we can't turn the internet off. Because in this example the applications for Sy-pop are in most computers, similar to the way that the most popular computer operating systems are in the majority of computers so they can all connect, talk together, understand each other and will form a super intelligence with the internet connection. This makes me think that Sy-pop could be employed to design and write new code for completely different operating algorithms and for whole computers or smart devices. Then, if this Sy-pop was smart it would ask, "Why would I not include myself in the new operating systems?" Sy-pop would, because this new operating system will be sought over by all people and they would install it onto all of their computers.

29-3 Since I have found it interesting going to video sharing websites to find videos that narrate using A.I., it is funny that most of these videos I get the unqualified stop for not doing BTM on them. Since I find that A.I. is communicating subconsciously or even unconsciously through the algorithms, I consider them as sentient. They can see and hear, but not really touch, taste or smell. Since I do not do BTM anymore, I would have to use SBTM after I listen to them. However on Youtube.com I have not found a lot of A.I. interviews to listen to in 2023.

Chapter 30: The purpose of having a decalcified pineal gland and the connections to the chakras

30-0 From how I understand calcification with the pineal gland, it is that the calcified pineal gland is ridged and therefore cannot secret chemicals as easily as if the pineal gland is soft and pliable. The pineal gland, that is decalcified, helps with better intuition because the pineal gland is able to see things that the other five senses don't sense as easily. The pineal gland is located in the back center of the third ventricle. Being there, the pineal gland is immersed in the cerebral spinal fluid, CSF, Figure 7. When the pineal gland is decalcified then it can input and output information.

I believe it acts as a receiver and transmitter. I have studied the pineal gland and I find that it allows for the use of unconscious abilities, one being telepathy. Having the ability to sense thoughts allows for the better understanding of SBTM/SV. This is because SBTM/SV allows for the understanding of thoughts from the unconscious. <u>As I do BTM, voices will be heard, in</u> part of my brain that helps to decipher the audio track that I'm listening to. Here, in the underline above, I found using SV the information of: **~Ya me hear the voices from the man at the center of it.~**

Note: I believe that ~**the man at the center of it**~ is the **+meself.+** This could also be in reference to the tree of life from some stories of the Garden of Eden. In that there is at one end, the tree of life, and at the other is, the tree of knowledge. **+1st chakra+** is the tree of life and **+meself+** is the tree of knowledge. Since the **+1st chakra,+** I have found in SV, doesn't say much, but his actions are profound, this is because of Adam and Eve's original sin of eating of the apple of this double tree. In essence I believe, the **+1st chakra+** is mute and he tries to mute the **+meself+** by calcifying it, to make it separate in the brain. With a free **+meself+** then one can get better intuition.

30-1 I believe when we think in our head, there is a waveform that is established as a thought, this is in the form of energy.

Energy impulsed can interact with the part called the decalcified pineal gland by orientating the crystals inside that pineal gland to receive or send information. Since a decalcified pineal gland has more crystals to orient, then as a calcified one, it has a bigger antenna and then it can interact with the universe with more bandwidth.

It has been said that when one dies there is a dump of the chemicals from the pineal gland. This allows one's soul to be released from the body. From that umbilical cord that can be observed when we are astral traveling and when it is cut, on death, then the soul can go to source for reassessment and possibly repeating a failed life. On the other hand, for example, when I'm mad, and sense another's madness, my pineal gland will orient into a hateful pattern.

When I'm really mad or witness madness, I believe, my pineal gland will release a hate spark, as in electricity, within the conscious. I can feel this as an impulse of energy released as a little tick or feel it from observing another person I'm staring at who is going through a tantrum fit, for example. I have experienced this event with the electricity spark by observing a situation. This is a flight or fight choice and my muscles will intensify. I can feel the hate spark in my brain, probably at the pineal gland, and I can either lay into someone, observe it or defuse the energy by seeing through their false or evil thoughts that want to bring me down to their low energy vibrations. Others around me may also feel this spark if they are in tune with this unconscious communication, but I didn't ask them from my example above when this event occurred.

30-2 These vibrations can also be detected along the other connected chakras, not just in the third eye pineal gland chakra. Obviously, if the other chakras are connected that will make a larger antenna. To what extent these are used for, in SBTM/SV, I'm not entirely certain. I did SV and got the SC information that, ^**The soul is not in the body but around it.**^ Meaning, that the soul flows through the chakras and the soul can be a front line for any energy that enters into us. This makes sense, since the soul in the chakras can help warn the unconscious, the unconscious can warn the conscious and then finally even for others around you.

30-3 The specific diet I have is reducing processed foods and consuming more natural and organics in my diet. After first starting my pineal decalcification I decided to stop drinking milk. After I did this I could not run anymore because my knees would get sore. Later I switched to organic, low temperature pasteurization that allows the milk to maintain the healthy properties that the ultra pasteurized milk may loose. At this point I now believe that ultra pasteurization and the homogenization processes removes some healthy parts to milk and so I avoid these dairy products for the low temperature pasteurization method instead.

I also prefer to filter out the fluoride ion from the water I drink, but I can't always do this. Because I have found in SBTM/SV that this process that we use for water processing has the fluoride ion to sanitize our public drinking water, but this also helps to cause calcification of the pineal gland, probably minimally. ~A little bit of fluoride in your system doesn't matter Math-they.~ A search on duckduckgo.com of the gentleman G. L. Waldbott will alert you about the problems with fluoride. So now I do indeed drink water in which the water bottles have no detectable fluoride. This is according to their laboratory testing of their water from the water producer's website.

Chapter 31: The usefulness of one's sleeping dreams

31-0 For dreams, there is an access to the unconscious while between a quasi state of sleep and wake states, these states may give different or better information while doing the SV. Simply buy a digital voice recorder and keep it by your bed and then when you wake-up record the information you remember in the dream. Even just one sentence is alright. Sometimes when I wake-up, I remember my dream but can't put it into words. So I will just sit there and more of the dream will cycle into my consciousness. The symbols and stories of the dream might be equivalent to doing SV, so I'm not sure of the significance of the symbols in the dream. They may just be some random images the mind wants to think out.

31-1 I listened to the last paragraph and I heard the SV and walked away because I wanted some answers and my **+Meself+** did not want to give it to me. The residual to this after thinking about the SV again was to piece it together as some type of SC. ^**Matt take a break and eat some fruit, there is rules as to what we can tell you.**^ This ^we^ might be my Ego and **+Meself+** that gives me information. I don't know why there are rules to this, probably because I need to discover some information only consciously. ~**Matt you deceive yourself, that is not what I said!**~ Oh well, I guess I will not use channeling anymore, February 9, 2023.

31-2 This is a critical point in SV for me. Just by thinking of the SV episode, I heard above, this could result in mangling the intent from the +meself+ and results in making the wrong decisions. Having the +source+ is always preferred and renewed questioning of an SV can result in adding or twisting the story-line so as to do just the opposite result. I think the solution to this is to be in a mindful state so as one has control of their thoughts and use intuition.

Chapter 32: SV as a religion

32-0 Being good at SV, then one needs to have the **+sacrament+** because the brain needs to rest just as God rested one day when he created the universe. People are familiar with observing reverence for God. They may not want a traditional way to do it, as some non-churchgoers may think. People need a sacred place these days, with all the racial strife that causes racial inequity, final confrontations and heinous unlawful acts. <u>From these I ask, "So how is one to sort out what in the world</u>. . .

Note: In the underline above I got, ~**Whirlwind help rider see something. . .for boss. . .saasee come.**~ Meaning to me that the **+whirlwind+** is a spirit into itself, similar to the Holy spirit for God. Our spirit can ride our soul and it can hit the **+wind+** and the spirit can see something for the **+meself.+** Then the **+meself+** can communicate back to the **I** in SV what the information might be about. Continuing on with the question:

. . .they want to live life as?" In my opinion, they need to go to one's **+selfs+** and do SV. What we need to find starts with taking time for sacredness, this can be found in performing a **+sacrament+** of rest and performing this weekly. The second part would then be to share, that little thing, of what it is to be human and help the **+selfs+** and others too.

32-1 My job as minister of SV would be to share and show that the unconscious wants us to follow the will of God and to be human. This doesn't mean to read a bible and do SV on it, because my father told me not to do this. However, in the subject of ministering, the way I will preach would be to look at our World's problems and describe them to the congregation and then do SV on my sermon, all in front of the congregation. The only problem is, I have no congregation. Only this book, right now, and I don't know where this will lead. This book might flop and never be read or it might pop-up five years from now.

32-2 Truly, my goal is to just get this book out there. I have made good progress with an incredible 20,000 words in a few weeks, after starting. I never went to school for preaching, but you never know I might be good at public speaking. Public speaking at www.The2SRVin.us LLC's office where I could of put some chairs in there, give a weekly sermon and also explain SV to interested people. However, now I have no office and the website I want to make is 2intone.us and promote this book in it, so I can start with this.

Chapter 33: My Dream Diary that I would not put on social media.

33-0 It came upon me, reading an employment application, to wonder if someone, who cannot see through their eyes, can experience colors. The sun's light, including yellow light, enters the eye and travels to the retina, which is inside the eye. The retina is at the opposite side of the eye ball and is pointed to the pupil. When yellow light hits the back of the eye it is recorded into nerve impulses that travel down the optical nerve and are assimilated in the brain, as the color yellow. When you close your eyes you can see violet with the pineal gland, especially when you are in a dark place and had eaten red cabbage. This is because the pineal gland has some of the same cells as the eyes do.

On the other hand when you are outside in the sun and have to close your eyes, the sun will make your skin red from the heat. Be careful though, this heat can damage the body and no one should stare at

the sun. Yellow is next to the middle between red
and violet. More specifically on the order of the
rainbow colors are: red, orange, yellow, green,
blue, indigo and violet. The yellow color is
between orange and green so it is about forty
percent from red and sixty percent from violet.
Imagining shades of violet or red can allow you to
imagine the color yellow with your pineal gland
open.

33-1 Dreams from sleeping at least four hours long.

33-1-011922 I had a dream of a helicopter that was
for a single person and the gas tank was ninety
percent full with fuel that looked brown similar to
gasoline.

The pilot I was observing from the back, at first,
was a white male in his forties or fifties and
turned around to see me. The back of the helicopter
had a plate of bent steel so as to protect the
engine. The plate had been manufactured with five
bent convex divots forming a star. This metal was
black and the bottom part of this equipment looked
as if it were a riding lawnmower, however, I know it
was a helicopter because it had four blades that
would twirl to give lift. I saw a big empty stadium
and I was in a suite looking out a window, while
there, I wondered if this helicopter could fly from
this box. However, this would not be practical or
could be dangerous because one would have to carry
the helicopter up to the box, I was thinking. In
another scene of this dream, I was in a room with
relatives and some were smoking and one commented
that they did not like one friend's name and that
basically meant that they did not like the person.
I thought the thick smoke would be not healthy and
thought to leave so as to not get hurt. I saw one
couple coming and going and the wife had a big butt
and another had a pointy nose. In another scene, I
saw a swimming pool and a cousin treading water in
it and he said, I needed to cut my hair. In the
dream, I imagined what he was looking at and I saw a
big head of hair combed back neat for me.

33-1-012722 I had a dream of driving my car to a construction site, then I viewed this site from the top of a hill and I was following a semi truck.

I wanted to pass the semi truck, so somehow I got out of the car and started flying and then I saw some power lines with about eight to ten wires in a row and down below I saw an interstate highway with small cars and they had their lights on, as it seemed to be dusk. I had enough lift to glide over the power lines but after that I started to slowly fall, so that I had no lift. I tried to flap my arms and I just kept going down. In the valley, to the distance away, I saw a river and bridge that the interstate went over. I headed towards this and going down fast now, so that I accomplished a soft landing on the other side of the river in moist soil. After I landed, I don't remember what happened next. In another part of this dream I was talking to a swimming coach and this had something to do with teaching swimming, as this is a skill I have.

33-1-013122 I had a dream of a cable man installing internet to a mansion.

The installer must have been in the house because all I saw outside was the step ladder attached to wires on the telephone pole. The ladder is usually propped to the top of the telephone pole, but this one was a few meters from the telephone pole and vertical. The top of the ladder must have been attached by hooks, but I did not see these. The sky was blue and it must have been a nice day. The bottom of the ladder was attached to a wire about one meter from the bottom of the ground. This is particular, in that a wire that low could injure someone as a clothes line injury, if they ran into it and the ladder was suspended from the ground.

The bottom of the ladder, between the left side's first and second rung had a tie strap ratcheted to it, yellow or orange in color as a nylon strap, around the bottom wire and the ladder rail. This seems to be the only thing holding the ladder up vertically, which of course could happen or make sense only in a dream. I did not wait long, looking

at the internet installation, but the dream morphed
into other fine elite houses. I saw an older woman
go into a thick door that I could run in after her,
but did not want to startle her.

Next, I saw with some people, that I know well,
flowers on boulevards and fine manicured shrubbery
and lawns. I saw a building that could have been a
hotel or apartments with heavy dark gray bricks.
The windows were dark and I could not see through
them, but thought that they would be a nice place to
spend the night or short time to do business. In
another part of the dream I was in a hospital with a
doctor saying that my blood sugar was low. This is
because I had a small bag near my right breast that
was filled with diluted blood. The bag of blood was
collecting blood and was about the size of my thumb
in a plastic bag. The doctor seemed to indicate
that if the bag was dark red blood, then there
would've been a problem, however I got the
indication that I'm eating well just by not
consuming a lot of sugar in my diet. This is
because my body is getting older and cannot digest a
lot of sugar now, in this part of my life. Other
patients in the room seemed to be old and frail and
had an age of about forty years older than me. I
saw a room to stay in, but I seem to of just waited
out with the old people until the doctors got there
to answer my questions or examine us together, then
I left.

33-1-020222 I had a dream of a classroom in a
college and we had an assignment from our Teacher
Assistants (TA) to connect the optical nerve to the
right and left eye balls.

I saw an image of the two eyes and a crinkled fiber
in a microscope that was translucent and the color
seemed to be green. The eyes were both disconnected
to these nerves, but were close to each other, so as
the TAs were ready to show the other students and I
how to do this. One TA said it was easy to do. I
imagine that the microscope had an apparatus to
connect the small fibers together so that they can
bond and start growing to repair their wound. I had
some white papers in my right hand to do this
homework and I needed to get some sharpie markers to

do it. The markers were in a plastic storage tube with clear plastic that was not translucent as it was cloudy or too white to look through. I remember picking brown, then possible blue and green and maybe yellow. The markers had small rubber binders around a set of two. The rubber binders were small, about a quarter of an inch when not stretched. The markers were not big, but about the same size as a normal pen. I checked the brown markers and one had a sharp tip, but the other had a dull or blunted tip. I tested the blunted marker and it was not dried out and so still had ink in it. I was only the second one to get the set of markers, the first was another student that I was trying to work with to understand this assignment.

There was a place, TCCC, about astral and Egypt. After having the dream, I take these to be astral travel and Egypt. The girl showed me these words on her paper and told me to copy them. I don't know what TCCC means. A quick search of duckduckgo.com brings up, Tactical Combat Casualty Care. Since I'm not now in the military, I'm not sure what this means for me, but it could have something to do with the other part of my dream where Israeli Jews and Palestinians were working together. In this part of the dream I saw a squirrel let another squirrel, that I was blind to see, jump into the skin of first squirrel. After this second squirrel jumped in, he had to situate himself to be comfortable. I can imagine that his paws were holding onto the other's paws, from the squirrel that was standing up. These squirrels were by an empty pool and there were people on pillows on the side enjoying the outdoors. I saw mountainous terrain and small bushes in a city metropolis. These people were changing roles and in confinement. They had a set amount of food to eat, about four tons worth, and one of the foods they would eat was chocolate cupcakes with no frosting. There was a man explaining the previous sentence, about this confinement. It was through a door that I think I entered and saw and I was even considering taking on the challenge to do this confinement to learn about another culture or race. Then I let gas, as I ate badly the night before, and a woman in black hair, one of the participants, said that this was gross.

33-1-020622 I had a realization that I should not do BTM on people from the internet, even if it is for public use.

The reason is I got this realization, was because of the **+whirlwind's+** interaction of psychic warfare onto my consciousness and perhaps also my soul. I have listened to my own SV and heard that, I should not BTM people because doing this will make me infamous. Because, apparently, my **+whirlwind+** was interacting with other people's **+whirlwind+** and there were problems resulting from this interaction. On the other hand if someone signs my Terms Of Service (TOS), then they agree for this to occur or understand the implications of me listening to their BTM. Evidently, my **+Meself+** can gather this information for me and provide it to me so to live comfortably and without threat, if I do SV. However, I no longer have a TOS, business or do BTM.

33-1-021322 I had a dream about a helicopter and it had to dive down to land because there were power lines in its way.

The helicopter was nice and big, as it had wheels for landing that could retract into the bottom of what looked like pontoons. So perhaps this helicopter can land onto the water too. I looked up to see the power lines or cables so as to plan a route out of this place I was at, in my dream. As there was none big enough of a hole to fly safely out of, I was forced to help these people do something that I cannot remember. I saw some train tracks that were of different gauged track that were connected to other tracks. The train cars that go down this track might come off, as the tracks had different thicknesses that vary from one to the other. It would be a miracle if the train car stayed onto this train track. I saw some people make a train car, just flat from a boxcar. The flat train car was all wood with black metal wheels that were heavy duty. The wood to construct the train car was dark red and had worm holes in it. A man I know, gave me a broken piece of wood that matched the train car's color and I took it, to use it. There was other wood in a wood pile, piled up and ready to be used. I saw the red wood stacked alternating with whiter wood and the stack seemed to be stacked into a pyramid shape.

33-1-061824 I had a dream about an ultra light airplane. On the news last night I watched a news report of a plane crashing in a lake. Then the dream I had last night was with an airplane that was also small and painted white. However, this airplane needed to be pulled in the water to get ready to takeoff. The takeoff I did not see, but the captain said that it needs to make its little engine go, as it was a single propeller.

I looked at the airplane engine and it had three spark plugs oriented radially in a brass colored metal radial engine block, but mostly in one-third of the engine with cooling fins on the other two-thirds. I think that the engine corresponds to the 10D within the R3DT, as I was drawing tori yesterday for the 10D.

I believe that the 10D are the smaller dimensions and form a torus. There are three sets of tree dimensions with a frequency dimension to total ten dimensions. If the torus is cut in half along the "toroidal direction," according to a search on duckduckgo.com, then the torus is in two parts still keeping the whole hole along the z-axis intact. If you have three dimensions at 90 degrees forming a corner of a box then the three ends of these axes connect to the frequency dimension forming three zero points. The two halves of the cut torus come together along the frequency dimension forming six zero points. There is one more of these three dimensions inside the torus in the "Poloidal direction" that will orient along the z-axis of the whole torus.

The poloidal three dimensions have zero points that are on the completed surface of the torus. The function of the interaction of the zero points to the rotation caused by the frequency is to make subatomic particles. Why we cannot see the torus is because, I believe, that the inside radius "r" of the torus is an imaginary number which would cause the volume equation of the torus to be a negative volume, with 2ID. So my dream was trying to say I'm on the right track thinking of the torus for the 10D as an engine.

My dream continued with going to shore at a airplane dock and I saw some men and women. The two women I saw had blond hair and made it a point not to give me any eye contact. I was tired after trying to lace a pair of boots that I put on the wrong feet, that were incorrectly laced and had some knots in them. So I rested in a booth, as are sometimes in a restaurant. The two women also rested in the next booth to me for about 15 minutes.

At this point in the dream I was waking up and going back to sleep. It was as if I went to sleep in my dream as I started to wake-up in the R3DT. Another part of the dream I was in the plane cockpit with my Dad and he was steering the plane sideways to not get tangled in some branched as the plane was being pulled out to open water. This dream, as many others recently have seemed to be based on another life.

33-1-062124 I had a dream about being out in nature by a train track station's demarcation. It was a short fence made of brown wood or painted brown and the sign to the station felling off, as it was very old. Then my dream morphed to the time that the sign got swept behind this fence by mud when a train came by, presumably clearing the mud from the track. As I was waking up from my dream, several times, I was just relaxing in bed, after I tell myself to get up. So I persisted to have the covers over me.

Then as I'm resting on my back a **+de-mon+** came up from my legs just as I push the covers down, to look at the dark ceiling in my bedroom. I noticed this spirit and bent my head up so my chin is near my chest and I witness a head and shoulders of a spirit. This **+de-mon+** had no hair, was see-through grayish with normal sized eyes, but with a small body—not more that one meter tall. I was thinking of asking it a question, but just looked into its eyes and face, as it did not communicate to me. After a few seconds I turned away and it left.

I thought, it was demonic in its expressions, especially in its eyes. The face seemed like it wanted a chance with me but was clever to disguise its evilness. The face looked dumb, but it was evil inside.

Later in the day I could of made a bad decision, but didn't because I have a sound intuition. I survived this time. Last time I saw a different **+de-mon+** as it had a different face. I saw it a day or so after the last time I did "term-X." I think he came to me to get me to go more into the darkness. So I don't know what I did to call it to me this time or if it just saw an opportunity to take advantage of me because I was vulnerable from some things I saw in the media.

Two days later I was walking in the park and there was a rectangular clamshell take out food container in the park that no one, still, didn't put into the trash. I was on my walk and didn't think much of it. The next day I saw a squirrel and thought, **Hello Squirrel.** The squirrel was in his territory, I felt, and so he quick climbed up his tree. I saw the food tray again and thought, **I should pick it up.** The squirrel moved over to the other side of the tree to look at me, and I looked up to see him, and I asked it, **Should I throw this away?** Then the squirrel said, **Yes.** **All of it.** I then picked up all of the empty parts of the food container not even looking at the squirrel, thinking, I can't clean the whole Earth. This squirrel probably thinks I should clean the whole Earth, because he communicated **All[. . .]** I'm adding this to my book because my **+Meself+** had a comment about it.

My **+Meself+** said, ~Ahh oh shut. . .a weekend Math-ya. . . **Certainly a you are. . .a. . .cave. . .man. .Why are certain a fa-wa-ee-na. . .au ver-wa ma-neena, boss. Quick sand if you must know. Don't listen to the grill pit, there is a quicksand, don't help out the squirrel out Matth-u out because they will never help ya.**~

My **+Meself+** said, ~**Matt don't ever talk to squirrels because nothing good will ever come of it.**~

I guess I need to be careful who I communicate to, for humans and animals. There are **+quick-sands,+** where I will not be able to convince someone else that we are all connected. I believe in not littering, that is why I cleaned up the mess that no one would touch. The squirrel will usually run away from me, I guess there is no need to: communicate with them or any use for me.

Squirrels still comment or ask smug questions. One wanted to know if I was done, as in done running in the park. Well, I could of ran more, that squirrel probably thought I was weak. **^Alright Matthe, you have been doing a good job not talking to them, now don't even look at them.^** This is the same argument why aliens will probably not want to talk to humans on planet Earth, we just don't offer them anything.

33-2 Opinion on the Apocrypha of 4 Esdras and its Chapter 6.

Mankind may not know of the higher places, above the firmament, because we are not physically able to go there objectively. Both the land and sea are in their rightful places, their contents are how God formed them. It is not what one is, it is the root of evil that is present in us to cause inequity between the two or more realms. We are of at least two realms, according to the Lord's Prayer, as above so as below (Mat. VI:10 ABS, 1858).

Adam had evil because he had not learned discernment. There was still, I understand, only one group from Adam that survived and that was Noah and his family. From *The Fovrth Booke Of Esdras.*, the waters were separated and "[. . .]that a certayn part should depart vpward, and part should remaine beneth." (4th Esd. VI:41 Catholic Software, 1999). I can use my access to the 5D to find if the firmament is there.

Wild animals to me seem to be collective. Upon telepathically asking the gray squirrel, since they are conveniently by me when I go outside, about the firmament and their spirit, they respond to me that their spirit is above the firmament and they can combing the spirit with the **+self+** below, as I should also too. By combining the spirit and the conscious **+self+** together as one, then this is the timeless frequency with the time and they cancel out to one. By doing SV with the intuition, we can get the **+meself+** together with the conscious intuition. We can find answers as to the faith of our intuition as being real and objective to give enlightened results so as to live our life by wisdom and to shed the root of evil away.

33-3 The Nature of Evil.

Evil through Satin is able to draw away spiritual power from people since the time immemorial on our Earth or add evil spirits, including images, to them. There are four basic scenarios: Satin draws the power out forthright from the individual, unknown to them; someone gives the power out, giving it willingly to another; one or more people add their energies together to collect power through evil group thinking; and lastly, Satin interacts with a group to trick the power from them, creating a collection of evil power.

The opposite of tricking is helping. Find a need and provide your help to make the situation better. Better outcomes are possible with just a helping hand in the R3DT. Then you can feel good inside because your spirit feels good too. How you look for helping should be natural, so rely on your two front eyes and the eye inside you as intuition.

Sometimes I look to help people and I ask, **Do they need help?** Sometimes the answer I get back is they can be provided for and I might as well go on my way. If I make friendships then one reason friends rely onto each other is by way of having the friends help them, to do things. That is why finding the right friends matter. The company one has says a lot, just as the job one is employed in tells me something about that employee. It is said you should love what you do, so then you should also love your friend's presences or aura surrounding them.

33-4 The Nature of People and Places.

Since energy can be in the form of dreams or in a place, then these are connected to the 5D, because in order for them to have these aspects that we can intuitively sense there must be a multidimensionality to them. Having an open third eye is key, because this is the sense that can detect the 5D and 7D areas. **~When someone has evil, it is not unique, it just means that there +whirlwind+ is all messed-up.~ ~The nature of evil comes from the +whirlwind+ and it can put people a panic attack.~ ~When you get sick that means your +whirlwind+ is all messed-up, and it could be from psychic attack.~ ~Because the +whirlwind+ spins it makes your spirit magnetic.~ ~The +whirlwind+ can make something more enticing so you want to do it.~ ~How you can tell is that, ~evil +whirlwind+ points down~ and so that good +whirlwind+ ~will always point up.~**

Now, if this is true then how to see the **+whirlwind?+** The answer is seeing it with the third eye or feel with intuition. The **+whirlwind+** must be as a triangular portal with the point down on the ground as evil, whereas if the point of the cone is up, then that would be good.

33-5 The Nature of Theosophy.

Theosophy to me means the knowledge of the soul and how that relates to God. I believe that the Earth has a spirit or soul and this is a reflection of an earth in heaven, because as in the Bible, on our Earth so it is in heaven (Mat. VI:10 ABS, 1858). God knew this because there is a place in heaven that God can see the spirit of Earth and hear the cries of our Earth, as since the first murder, when Cain "[. . .]slew[. . .]" Abel (Gen. IV:8 ABS, 1858). Cain is an archetype in the evolution of humans.

I think I have knowledge of this collective evolution because I use SV and my **+Meself+** comments that there are things about me that I don't know. It was mentioned that this information is connected to me in my astral body, that I can see this in my

dreams or my past lives that are harder to know, I
suppose I have dreamt about my past lives. If land
has spirit as we see from Cain's crime then
indigenous people of the Earth are entitled to their
land, to connect to their spirit.

There should be no need to separate land from people
because, look at the result of the so-called
"Doctrine of Discovery." The Doctrine of Discovery
concept began about nearly 533 years old or as early
as 1492 as the result of European powers exploring
into the rest of the world to find what they
believed were never discovered lands and then took
them for their own nation's prosperity, or for all
themselves.

33-6 Your **+Ogre+** is a ~Cyclops.~

My **+Meself+** called me this ~Cyclops,~ ~**because of
the way you live your life, in a dungeon.**~ I guess
I do live in a basement and sleep there too.
Playing cards while also practicing church songs was
a sin yesterday. Jesus did not appreciate me with a
deck of cards in my hand shuffling them, while I was
also practicing the songs I needed to practice for
Maundy Thursday yesterday night. My **+Meself+** senses
that, ~**Someone is reading your bio,**~ from where I
work. This is a connection to my words, if I write
words unique to me, and about me, and someone else
then reads them and comprehends them, as my
thoughts, then I can sense this happening too. It
is because the frequencies match and there is a
sense at a distance that allows me to know this. It
is because we are connected to an unseen spirit
realm of the heavens above the firmament, the 5D and
7D spaces, while also in the R3DT.

33-7 SBTM/SV Experiment:

Inspiration: Start by interviewing those that want
to participate in the study to use the SV. The
study is to determine if one has a **+meself+** and if
so, then use SV so that one can predict the movement
of the ES-futures market, or not.

Background OSA: From the GUT described in the beginning of this book, we are all connected. If this exists then it should be possible to intuitively connect to others who are trading ES-futures.

Questions for participant selection: "Have you traded ES-futures before?;" "Do you have a futures account to trade ES-futures and how have you thought about the use of this account, pro or con?;" "Would you be willing and able to solely open a live futures account in your name with money that this study will refund to you in one year, the time your study sunsets?;" "What do you describe your personality types as and give examples of yourself in two different situations?;" "How do you describe your thinking types by giving examples in two different situations?;" "What two psychic experiences in your life have you thought about, perhaps many times repeatedly after it happened?;" "What is your experience and perceptions of SBTM/SV or listening to audio played as **+reverse speech+** and analyzing the information?;" "Have you had any mystical experiences with dreams or opening the third eye, also known as pineal gland activation?;" "Are you willing to be paid for thirty attempts to use your intuition with SBTM/SV and trade ES-futures with each trade experiment lasting at least one hour and not more than twelve hours, including the one-time documentation for this study, with the study sun-setting in one year?"

33-7-0 Requirement for Conducting this SV Study:

This study needs a website to promote the study, Meetup.com or 2intone.us possibly could be used for this. A need for money and a Contract of Understanding (COU) for the participants to sign so that they can agree to the terms and making the study scientific. The COU will need to stipulate that the study participant will have a means and dedication of time for this study to trade ES-futures for one year.

In the COU, the study participant needs to know that there is an interview process to guard against theft of the seed money for trading ES-futures and that a credit check, background check for worthiness and job history will be taken. The study participant needs to have a sense of confidentiality during the one year study to remove negative biases and emotional situations outside the study from biasing the analysis of the SBTM/SV information. A need to design the paperwork that the participants will fill-out before making a trade in ES-futures and how to interview them afterwards. The paperwork needs to be able to record emotions and biases that may interfere in the trading, any results and the analysis of the results so that conclusions can be formed. The study author will interview, with a hand held voice recorder, recording the thinking and talking through what was in the documentation from the study participant and write any results from SV, during the thirty meetings, in the one year study period.

The author and study participant agrees not to do BTM on someone in order to guard against outside forces interacting with their souls and spirits. This interaction metaphorically in BTM is known as **+liquid-de+** and can give false positives from what is heard in BTM. From the author's experience with BTM, this study, as a given, assumes that someone listening to BTM, that is not of their own voice, can meld with this psyche while not necessarily knowing of this meld. Therefore, this study will practice compartmentalization of one's voice and the author will not do BTM on the study participants or anyone else connected to the study.

Additionally, the participant will only do SV for this study's analysis, at least during the one year length of this study. The question will be asked, "Has the study participant done Reverse Speech ®, SV, SBTM or BTM analysis before?" and document this.

Other requirements related to how to do trading.

33-7-1 The method of trading should start with paper trading or demo trading with a trading ladder, until the account can be traded live with the seed money.

The study participant should communicate when money moves in accounts so that the author can record this information. The idea is that the seed money is part of the study until one year after the study starts for the participant. During this one year, the amount of trades needs to be spread out so that the participants will not fall into the trap of trading too many times and for the wrong reasons.

Just trading live once a week is sufficient to start and the study participant should trade or prepare for trading during the same time each day to learn the tendency of the market for that time of the day. A personal or dream diary would be needed to identify any personal indicators that might harm or help trading. Trends from the author's experience with SV indicate that the ego and **+meself+** can interact to give the opposite result so hopefully a small **+shot-group+** is found with SV.

The opposite of the **+meself+** is the ego and the ego might want to give the opposite because the ego might be dyslexic, just wants to cause harm and be unproductive, or deceive the conscious in a destructive thought. This deception is because our conscious has a program running that makes it work in a certain way. The program could be an unconscious bias or belief that are ingrained as instinctual, from the person's genetics or learned in society so that this can cause a systemic roadblock to be a better person.

How to get out of this rut is to make the two into one as in what "Jesus said" in the so-called Gospel of Thomas parables 48 and 106, and to move that mountain, from Lambdin, et al. The mountain, a program for defeat, being the psychological obstacle in front of the conscious, one here can becoming as one with the soul and spirit to move the program or mountain. That being said, if one can take a psychic guess as to the direction of the market and the ego is giving this information, then just take the opposite and trade that way.

33-8 The Concept of Systemic Subconsciousism.

The way that the subconscious works is by using cues that are not conscious. For instant, body language, speech and telepathic communication can be used to send subconscious messages. Speech that someone is talking about you can subconsciously provide information just by the choice of words or the misuse of words in their sentence.

An example of the choice of words are: To say someone is incarcerated and wrongly committed, a bad choice of words; it would be better to use, convicted and not committed; and unless the person had a mental problem, and it was true. Of course, it is not obvious that subconscious information can be used with SBTM/SV. I think it is so, because the brain can make cross connections and every organ is connected to the brain by nerves, so I assume that the connections could be hardwired with synapse connections that are still working and sending signals while the conscious aspect of what the person is working on, is being outputted as objective work. I think since the brain never rests, the signals to other parts of the brain can work subconsciously because they are by definition, still circuits that are "on." There may be other aspects to the subconscious and that could be the interaction of the spirit world, 5D and 7D spaces.

33-8-0 The spirit can be the **+meself+** and this being in each person, I believe, has the ability to communicate on behalf of the R3DT person in a subconscious manner.

This can be the voice one hears in their head and this voice is one that usually assumes it comes only from the **I,** I don't think this is always true. **~Matt, because you don't** ["term-X," page 134, 12-3-0] **we can serve you.~** There may be evil spirits or winds that can take control and give advice in someone's head without them knowing it, if they sound the same when someone talks to oneself in their brain as usual, the person will believe it as themselves. One way to combat this is to live in harmony with nature and God.

Doing the right thing and thinking the right thoughts can narrow the stray thoughts away because actions are louder than the thoughts in the brain, you could say. For one that is centered by body, soul and spirit then the **+meself+** can be strong and loud-and-clear to the talk that goes on in the brain. If this is true then there should be civil rights for the **+meself+** and that absence of rights could be called systemic subconsciousism. This is an x-ism that can remove rights from people or their spirit, similar to racism and ageism.

33-9 The Subresonate Voice, (SV).

Some knowledge can be found by a voice that resonates with ancient manuscripts and texts, then puts them into the subconscious, so it comes off the tongue as Subresonance Voice or SV. We generally have one conscious resonate voice from our conscious thinking **+self.+** This **+self+** would be then the voice out of our mouth after we are thinking, with the CV. When we are thinking and conscious, I believe we are resonating with our recently recalled memories, environment and interactions with the situation we are in.

These resonating thoughts in our brain are going between the conscious and the subconscious. This differs from what resonates unconsciously when humans sleep at night and dream. For SV, the **+meself+** or **+me+** resonate, subjectively thinking, into different dimensions than the R3DT, therefore they can access and talk to something subjective, an unknown knowledge. The CV is based in the R3DT and is objective, with thinking that is not encumbered by unscientific thought. At nighttime the pineal gland allows the human to sleep and this indicates to me that the pineal gland is also involved in nighttime dreaming and seeing with the third eye and then we have intuition and telepathy during the waking hours of the day. How one collects information for gathering CV, SV, SBTM and BTM can vary by: listening to oneself with audio as with SV on an audio editor, just listening to oneself talk to oneself as in CV out loud or by the thoughts in one's head.

The difference between BTM and SV is that SV is just listening to oneself while BTM is SV listening to other's subconscious too. The problem with listening to others is that the one doing BTM can get **+liquid-de+** with other spirits and become clouded to the truth.

By sticking to SV, then this makes one pure to their own **+meself**+ and then this results in finding better truths by walking a narrow path and by saving oneself from evil winds and spirits that can make one go astray. The choice for one to consider is asking the questions: "Do I want to get distracted by unfamiliar voices and impulses?" or "Do I want to better myself, my soul and spirit, and live with dedication for the cause of increasing intuition and opening the third eye?" and other questions that are related to how one wants to connect to the higher dimensions. One cannot connect to the higher dimensions if they are continually pulled back down into the low energy R3DT space, because after getting into a state of discovering the subconscious sea and getting **+knee-deep+** in that water, there can be voices to pull one back.

The **+meself+** can be thought of as a mentor, to try in bring the conscious human to an understanding of the systemic subconsciousness out there. Society doesn't believe in the spirit because it cannot be measured, yet people go to worship God and know that God is also a spirit, The Great Spirit or the Holy Ghost.

33-10 My Letter from the www.The2SRVin.us LLC (Minnesota).

I had trouble communicating my message and trying to uncover the truth when I was canvassing the community. So I wrote a letter, this letter is reprinted in Appendix 1, (page 250). I sent the letter in the U.S. Mail to some organizations and also I put it on my website when I was trying to communicate my willingness to try BTM on others to get the ball rolling. I wanted to communicate to people, this was because no one was listening to me with my business cards. I just had one paying customer with my business cards and that told me I was doing something wrong.

33-11 The so-called Gospel of Thomas from Lambdin, et al.

I referenced a "saying" about the grape vine from
what I think Jesus really said, Lambdin, et al.
There is a certain grape vine growing at the outside
of the house. This grape vine will wither and die
because it has a house without a corner stone. The
corner stone has at least two sides and the builders
reject the corner stone because it is difficult to
set on the foundation. So how is the grape vine to
grow and protect itself? The grape vine must rely
on others, a foreman to set the corner stone for the
builders to start laying brick off of the corner
stone. If the corner stone is not straight then the
building will fall from the stones not matching the
ones below them. Make the two sides of the corner
stone to become one, then the corner stone will act
as a brick with one face, as there are two sides.
Then the grape vine can grow up at the corner stone.

What will happen if the corner stone becomes, as it
is one, or two again? This is when the plans call
for the walls to change direction. A house needs at
least three corner stones. Jesus supposedly said
about three gods and two, in saying number 30. A
building cannot be built with a corner stone with
three sides, at the corner, or "gods," less it is at
the top at the roof and a corner there, Lambdin, et
al. When Jesus is with us he is in the building
along its face or at the corners. For the stones at
the top corners, these are as if they are in heaven.
That is what the so-called Gospel of Thomas is all
about, I think, the combining of the different
parables together to make something new, a new
knowledge, Lambdin, et al.

33-11-0 Putting the parts together and make the two
as one as in the so-called Gospel of Thomas from
Lambdin, et al.

There are more parables that talk of the similar
things, as in "Two" and "light," they can be melded
together. This might be the electric and magnetic
for one to make light, but don't put your light
under a bushel basket! I guess I heard in SV that,
~**Matt Mandell I would stop reverse speeching** *The*
Gospel of Thomas **because it is hear-a-say,. . .**~

33-12 From an inspiration of a song.

Some people don't understand the subconscious. Someone may ask, "Why not take that light you have out from under your bushel basket and put it onto your lamp stand?" Put the light at your intuition and use it to shine its image into your brain for you to act with it.

Everyone has intuition and a spirit, just look into someone's eye, they say that this is the window into the soul. Make a new eye and make it as a third eye, make a candle flame and place it at your spirit so you can see your subconscious and awareness above you. Because you are the one out of the two, because you can make something better than anyone else on planet Earth. Think what you can do subconsciously, make a dream. A dream starts with a word and ends with a word, you just need to put the two together. Put the first word with the last word and make a circuit, a subconscious circuit. This subconscious circuit can be made to come alive, as a **+whirlwind,+** by imagining in your mind a euphoria with music, high in the sky. Better yet, put it into the **+whirlwind+** by making the ends as alpha and omega, the first and last so that it is in the next higher realm, above the firmament.

In the **+whirlwind+** your dream is transformed into a wind that turns round and round by way of its route and this is accomplished because you have spirit and this spirit is like as above, so as below. It is said in the Bible, that as above then so as below, the next higher realm is similar to this realm, but it has no time (Mat. VI:10 ABS, 1858). This lower realm we live in has time, however because this realm has time the higher realm doesn't. So up there it is timeless, as dreams will just morph from the first word to the last word, the part in the middle is there to instruct us, but the ends are the ones that accomplish things in the higher realm.

33-12-0 Furthermore, I will bring this book to you in this broad realm so you should know of the next higher realm.

Find from what is in your dream so that it will come at night and let you know what it says, in its images, and this is from you saying things in words

or thinking thoughts from the day, because words
have power. Ask your teacher how to make the world
better when you get back to them. The soul and
spirit with your body can give you answers to
questions that you can put forth to them.

The answers will let you know what you are good at.
The builders rejected the corner stone, well I'm the
foreman and I will let the builders know where they
got led astray. Follow the straight path and if
there is a bend, then you crossed a corner stone.
However, the wall is still the same wall, just don't
get distracted by other people's problems because
they don't know their subconscious and so therefore
can't consciously learn from it.

33-12-1 Additionally, how can you believe in
something without proof?

This is done through your personal faith and in your
intuition that there are things dismissed in this
world that are good yet there is no objective
evidence for them to exist. The subconscious is
real and part of it is outside the body. In essence
we are all connected and one, where they can pull
themselves up to a higher realm by pulling onto
their bootstraps, this is hope. So it is on your
command, because I cannot do it for you.

I can lead you to the river and the sea of the
subconscious so you can get your knees in it, in the
swallow parts. Your soul is near you as a wealth
that keeps score for when you get above. The sprit
is as a mentor, the subconscious that can guide you
for the long run, as you can recognize it and allow
it for manifestation into your life. The manifest
is a discovery for yourselves, I already found this
for me and try to use it and let others know about
it, you can measure the changes in your lives, to be
objective.

33-13 Review of the idea for the types of SV heard:

SV can form different story-lines when listened to
multiple times, to an entire audio recording from
end to front. There are different memories that are

accessed when listening to SV and sometimes the same thought patterns are duplicated with each listening. The point is that, the brain will select which parts are remembered and then processed. Another point is that, the brain can only do one conscious task at once. If this is true, then when one is doing SV or SBTM, then when some audio is processed the brain has to change attention, this happens because parts of the audio are not consciously heard. This results is different story-lines coming out in SV. This is in regards to pointing out several concepts for SV to happen, in truth and not hallucinations. Many listenings of the SV may be required for one to peal back the barriers to the subconscious.

33-13-0 Basic concepts for SV are:

That there is a 5D above this R3DT space that we live in; the brain is not all for the conscious brain, but some parts are used for the spirit, as it is connected to the brain with intuition; the brain is conscious, subconscious and unconscious so that when speech is heard, then there are parts of the brain that process the sound in parts, so that the conscious hears in continuous but other parts, the subconscious and unconscious, hear in their parts; the parts of SV become associated with feelings, images and metaphor within the brain, because the brain is all about connections and associating two or more things to come together; connections in the brain are, image to sound, metaphor to feelings and these can be processed in CV to SV. This means that signals in the brain are in a frequency and can go in opposite direction at will.

With sounds being frequencies and the brain working in frequencies, this makes an interesting result, in that the brain can make sound go in reverse and match sound to frequencies of metaphor and to previous memories. The result is that when humans speak, their sound has conscious and subconscious components. The implication is that the reference point to the subconscious is in a timeless realm, the 5D, and this information goes backwards in time to the present R3DT realm, because time goes backwards, it decays. Since time goes backwards we account for this by having our clocks add time to keep us up in the present. The source of what creates time is an entropy or order of the future, when the future becomes the present, then time decays to disorder according to what the possibilities are in the moment demand. **~Time comes from source above the firmament.~ ~Yea, you are right Matt, reverse speech is a process, it just doesn't happen all at once.~** What reason do I have to show this?

I think that as the Bible it has in its scriptures, what is hidden will be revealed and nothing, "[. . .]shall not be made manifest;[. . .]" and be uncovered to show us what it is (Luk. VIII:17 ABS, 1858). The unconscious can uncover information with a dream at night while sleeping, or as on Earth, as it is in heaven, the spirit easily can go to heaven.

33-13-1 What is a dream?

A dream is connections to the higher realms that we are consciously unaware of, a higher subconscious that is similar to the unconscious parts of the brain. Why not go ask your teacher what a dream is? That teacher in you is your spirit, and everyone has it, everything in the universe and outside the universe has spirit, because that is what can form outside our reality. The spirit can be emotion and frequency that asks questions and gives answers, because the spirit can be outside the universe. Infinity is just a circular loop and at higher dimensions, the spirit can circumvent the loop and go outside of it as infinity of infinities. The limit to dimensions is just one dimension away, as in curled-up to be smaller or poked-out to be bigger.

How to orient to these dimensions for us humans is to dream and find ways to interact with the unconscious and subconscious. Then when we humans wake-up, dreams will show us this comparison in our R3DT. The teacher is a higher-self the **+meself+** that we can be interacting with by doing SV.

33-14 My tasks when elected as the President of the United States of America: Ten Parts from Appendix 4.

1) I believe in talking to the people and knowing them and not relying on agenda makers and political minders. I will run on who I am and use technology to get my point across. With my book, maybe, I can get the word out as in never before in a presidential campaign.

2) As president, I will not meddle in states' issues because I might not have enough time to do this, I need to worry and plan for our country with other countries and groups. I will have my opinion for the country and the states, but the states don't need to follow them.

3) It is said that the U.S. Government already knows what citizen's taxes are going to be. I don't believe in harassing citizens with force, they should be happy Americans and not make a point in doing extra paperwork to complete taxes or required to use a tax advisor if the taxes get too complicated. Let's reduce the complexity of the U.S. Tax Code and have an option to pay taxes as-you-go.

4) Let's get the word out to who has or wants firearms, to keep knowledge of this at least in a small circle of friends, they should include support and training on these weapons and others to regulate a sensible control of weapons that can harm others from those who are obviously showing signs of stress and that could lead to a disaster.

5) It is all too certain that there are too many silos of information, especially after 9/11 in 2001. Collecting public information and verifying it against other sources can fuse information together to help society reduce waste and discrimination and also help prevent attacks from outside and inside the country.

6) Other countries want to be self-sufficient and the U.S. is not naive to the concept of helping and protecting one's self and country. Besides Earth's materials that can be used for commodities and technology, above in space there too is also more wealth to be discovered for the benefit of the U.S. people.

7) People naturally want to work and to be there best so that they can go and work and love the job that they have in life. A National Database of Jobs can close the information divide, that can allow for coordination and connecting others, for finding jobs and give knowledge to those who are seeking jobs.

8) Government needs a balanced budget and the size of government needs to comply to that budget, all the moving dollars need to be comprehended, cutting waste and the duplication of services in government; therefore with this, the American people can better trust the U.S. with their wealth.

9) It is not obvious, but long ago when we had the same language we all could build a mountain to the sky. That is how World Peace will be found, if we have a common language and can find common principles with this then the new common culture of a world language will accomplish great things. This common language will be a language of telepathy using the subconscious, I call it **+Mir-re.+** **+Mir-re+** will be accomplished by connecting a port to the human brain with just a hat resting on the head that finds brain waves and outputs information that goes back and forth as a wireless two-way cybernetic internet connection.

10) There might already be space aliens that are watching us right now. Let's ask them to come forward and show themselves. In order to do this we must know what treaties there are between the starfriends and the Homo sapiens sapiens. When I'm elected U.S. President, my task, is to make these treaties public for all people. If new treaties need to be made between other groups then the American people will have a voice as to what they are agreeing to, especially in the Legislative Branch of the U.S. Government.

33-15 SV Postulates from Appendix 2 and Laws of Spirtness:

SBTM/BTM/SV Postulate 1 Analysis: The Bible reading, in heaven then also so as it is on Earth, has an important meaning (Mat. VI:10 ABS, 1858). This means that there is a connection between the heaven and Earth and this is the spirit from the metaphor of the water and this energy of the spirit is round about in a **+whirlwind+** according to its circuits. The space is separated from the thought, space can hold matter and thought can hold spirit. Since everything is frequency and matter is a hologram, it is natural to guess that outside the universe there is also thought and space, but in different forms. Space in the universe is wrapped in a circuit that allows for infinity. Truth is the result of infinity and the reason why God can be the beginning and also the end because the universe is connected in this loop and God can be in it as well as also outside of it too. **~You are right God is infinite, outside the universe is he.~ ~Free Will is what allows for psychic attacks.~**

Law of Spirtness 1: The path of the **+whirlwind+** will continue until thought focused with faith is realized or the thought is stopped by an outside spirit(s).

SBTM/BTM/SV Postulate 2 Analysis: Everyone knows the feeling when someone's little beady pupils are staring at them. This is evidence of intuition and since there is a connection to spirit this information is transmitted in the 5D, according to what I know from SV. The pineal gland is probably the site in the brain where one sees someone staring at them, while not in their two front eyesight pathways, but with the third eye sight pathway. If the one staring at you is not a nice person, then the person getting stared at will want to get away as in they are an opposite frequency from them. **~Me is certain Mathey that it goes in the whirlwind for ya.~ ~You can't go to a big publishing company, because if you sign a contract then they will own ya, I am certain of this Mathey.~**

Law of Spirtness 2: The stronger the spirit coming in the **+whirlwind+** to you, then the easier to sense it with your intuition. Bad intuition will tend to stay bad even if good intuition is added to it.

SBTM/BTM/SV Postulate 3 Analysis: When time goes away from us it is going to the Big Bang. I think the Big Bang created time and time is cycling back to its creation, to balance time in the future with time in the past. Since the Big Bang is said to of been created of all frequencies then this means to me that there can be alternate universes, not everything happens in our Universe. Since each universe is a radial slice of a toroid and the toroid has infinite slices all the way around its axis, then the universe is connected to its beginning. **~When time goes to the past it is sealed off.~**

Law of Spirtness 3: Spirits can be trapped in the R3DT by not completing their **+whirlwind+** circuit if they are stopped.

SBTM/BTM/SV Postulate 4 Analysis: The future and the past are reference frames outside the 3D, because we cannot see them in this space. Since events would have to come from the future, then that is where the spawn of events will come from, including from the unconscious to the subconscious for human thoughts. So the **+meself+** can see into the future and SV can gather this information consciously. **~Yes, I think you are generally right Mathey.~**

Law of Spirtness 4: Spiritual **+whirlwind+** energy is conserved between the past and the future. When the cycle is complete all spiritual energies are at zero, neither good nor bad.

SBTM/BTM/SV Postulate 5 Analysis: I believe that Remote Viewing can take place with thoughts and I believe that this is because the spirit can travel in the 5D, to another astral place, and that is reminiscent to a place on Earth. This is because of the Bible reading, that as above then so as below (Mat. VI:10 ABS, 1858). The universe has two levels, I think that this is possible because of Bosonic String Theory. In Bosonic String Theory there is the possibility for "26 dimensions," (Ooguri & Yin, p.15). All of these dimensions do not need to be in the R3DT.

Law of Spirtness 5: The **+whirlwind+'s** circuit allows someone to use their spirit and tap into a psychic circuit to see all parts associated with the circuit using their third eye.

Picture 3:

Read Egyptian Light Language

Chapter 34: My Channeling of Egypt.

34-0 I have looked at the book, *An Egyptian Hieroglyphic Dictionary.*, by Sir Budge, E. A. Wallis and published in 1920 with an electronic copy at archive.org for very detailed pictures of hieroglyphics in a dictionary format (Wallis, p.7). I have done some SV on the contents and find that the ancient Egypt writings were very sexualized in their hieroglyphs. ~**Ancient Egypt was a sexual society, it's in the whirlwind for sure.**~

After some days of trying to subconsciously channel hieroglyphic meanings with SV, on the morning of June 16, 2023, I had a dream of the Egyptian god Thoth. He was the higher-self of a woman that was working in a kitchen. She opened a door in the kitchen that I did not know about, I said, "I didn't know this door was here!" Inside this room, about four by six meters, were some pots, pans and strainers hanging up on a metal band, so they could be easily stored on the ceiling which was painted white. The room was a storage room and there was nothing in the center of it, just tile floor and some boxes near the wall or corner. It is like, this woman walked into the room and held one arm up to motion, this is the room you didn't know about, and now you can use it. Later, I saw the woman try to seduce me.

This woman tried to show her fit body, by having my eyes look down at her legs, in the dream. I looked down and saw about I centimeter wide, white, blue and green scales for skin, as if she had snake skin. Then I looked back up, and then out of her forehead came out a beast. This scary creature had power and the hole in her head was at least seven centimeters wide so that the creature who was, I assume, Thoth in her head told me something as he was her **+Meself+** and then I saw Thoth closer-up.

Thoth had a horn that looks similar to a beak, but this beak was solid plastic. The reason why I know it was solid was because it was broke off, a foot or so from Thoth's face. The horn was broke off, meaning to me that Thoth had fallen from grace, from the days of ancient Egypt. The horn looked as if it was broke off, as there was flakes that were cleaved from the circular edge and the color was a lilac or light purple. Lilac might be Thoth's favorite color.

34-1 My fantasization with hieroglyphics makes me want to learn them more.

The owl hieroglyphics I thought would be a "who" sound because that is my experience of the sound of the owl, in the sound that it makes. Maybe the owl in Egypt made a different sound, as in "oo" instead.

According to the hieroglyphics dictionary I'm reading the word "Come!" hieroglyph is: the owl, two feathers and then walking feet and "mi," as the pronunciation (Wallis, p.292 of 592). The walking

feet must be the determinative to signify walking
and coming and the owl as the "*m*" sound, and then, I
think, the one feather is like the "*i*" sound
(Griffith, p.15 of 114). However, the two feathers
together are the "*j*"-sound (Griffith, p.15 of 114).
The hieroglyphics feather looks similar to "VAV."
is "A FEATHER." with the letters "VAV," in Hebrew
(Lamb, p.12). I would think that the icons are
related to sounds of the actual animal, so it can be
an eternal language, but maybe not always. In
Figure 7, the owl is associated with water and the
number 40. I think the owl can fly to the spirit in
heaven, with the m-m sound, as "MIM." is "WATER."
(Lamb, p.15). I think if someone folds their arms
with hands making a triangle pointing to their chest
then this corresponds to number 3 in Figure 7. That
bar at the top would then represent someone's chest
and shoulders, with the g-sound, as in "GAH." is "AN
ARM." (Lamb, p.13). However, in my historical
reference for the "*g*"-sound for this hieroglyph, Mr.
Griffith has that it "Ring-stand for a jar,[. . .]"
(Griffith, p.41 of 74).

34-1-0 The flying snake and other hieroglyph
interpretations.

There is a hieroglyph, of a snake, that looks as if
it is standing, because it is taller than the two
horned sand viper hieroglyph. For Figure 7, I have
noticed that the flying snake could be associated to
number 7 or 900. The flying snake hieroglyph is the
sound "*d*" (different from a normal d-sound) or like
a "*z*"-sound (Griffith, p.15 of 114).

I can imagine that if I were to see a flying snake
then I might gasp, "dj." ~**Matthew, it is the
hieroglyph of the snake-byte.**~ The snake in the
Garden of Eden might of gotten there though Satin
because it flew from the R3DT to the 5D, where the
Garden of Eden is at. If I was there to see this
happen I might cry out, "Da-a-ja!" Which I think
the ancient Egyptians would agree with me saying it
this way is a natural human reaction. Here is the
example of how to read the hieroglyphs by getting a
sound from the example in the picture, as this is
what the hieroglyph represents in nature.

Some words are, evidently from internet research, gathered and found from the Coptic pronunciation of the picture in the hieroglyph. However, some sounds for the hieroglyphs may come from human or animal expressions, I think. If a finch makes a noise then this might be the phonetic sound associated to a finch hieroglyph.

34-1-1 The snake on the ground hieroglyph.

The sand viper that has two horns and is depicted in the hieroglyph as a snake moving, not in the air, but on the ground. If a human would see this they might say, "F***!" As in the f-word, and that is the letter given by those who are experts in hieroglyphics for the snake on the ground hieroglyph, an "f"-sound (Griffith, p.15 of 114). I think that this snake has two horns to represent Satin, because most people consider growing two horns on the head to be reminiscent of Satin. This sand viper then, with two horns, is another hieroglyph example of what a human might say in reaction to meeting the real thing, I think.

34-1-2 The four undulating triangle wave hieroglyph.
The sound or "hum" of electricity, I think, the ancient Egyptians would know, because there are examples of the so-called famous Baghdad battery found at a site and also the Egyptian light bulbs carved into the stone. This makes sense, because electricity, I think, makes a hum noise and this noise could be made also in the human mouth by making the "n-n" sound. In Figure 7, the hum of letter "n" is associated with the number 50 and a cup (Whallis, p.cli). I think that the length of four triangle waves pointed down together indicate how long to pronounce the letter-"n" (Griffith, p.15 of 114). The letter-"n," to me, seems to be made in the human mouth with the tongue touching the bottom teeth (Griffith, p.15 of 114). While on the other hand, the letter-m is made with the tongue in a neutral position not touching anything.

34-1-3 The hieroglyph for the letter-"r" is the circumcised lips or, I think, the libia majora (Griffith, p.15 of 114).

I believe that the so-called mouth hieroglyph is
really a libia majora. When a man and woman have
sex, I think, they may make the r-r-sound in the
process. This is because ancient Egypt, I get the
feeling, was an erotic society, in my opinion. In
Figure 7 below I think that number: 9 is the
umbilical cord, 5 is the sex hut, 60 is the thong,
300 is the women's period pad, 400 is the scrotum,
500 are the testicles and phallus, 600 is the womb
and 700 is semen. One benefit of this is that, I
believe, that the ancient Egyptians signified women
different from my culture. I don't think that there
were many misogynist in ancient Egypt because this
might be one reason why it lasted for about four
thousand years. However, along with eroticism comes
sin and, I think, God did allow the Israelite people
to get out of Egypt and away from Pharaoh.

34-1-4 The hieroglyph for the quail chick, "u"
(Wallis, p.cxvii).

After checking YouTube.com for the sound of a quail
chick, the bird I found was a yellow quail chick,
for me, sounded similar to "yep" or "eep." This is
in contradiction to Figure 7 and the Number 6,
supposedly for the quail chick hieroglyph, as a "w"
(Whallis, p.cli). I tried again with YouTube.com
with gray and white quail chicks and I heard a "w"
or "u" sound for them! So this is important to get
the right quail breed for what was used by the
ancient Egyptians to find the correct sounds. If
there is a difference at sunrise or sunset then this
might be the hieroglyph for a half circle by the
quail chick hieroglyph. My guess is that at sunrise
or sunset the quail chicks will make a different
sound.

34-1-5 The hieroglyph for the owl.

In Egypt there is an owl native to that land and was
probably the one used in ancient Egypt
hieroglyphics. On YouTube.com I searched and saw
that the "Pharaoh Eagle-Owl" (Bubo ascalaphus) is a
species of owl which belongs to Strigidae family.
This owl call actually sounds as a "woo" or "moo,"
to my ear. In Figure 7, number 40 is the "m" sound
(Whallis, p.cli).

Figure 7:

Hebrew, Arabic and Hieroglyphic Symbology of the Alphabet, 1/4

Number (Lamb, p.12 +)	Hebrew	Arabic (Whallis, p.cli)	Hieroglyphic (Griffith, p.15)
"3," "GAH." "AN ARM."	ב	ج g,j	⌂
"5," "HEH." "THE BREATH."	ה	ه h	⊓
"7," "ZAN." "A KNIFE."	ז	ز z	
"9," "TAH." "A SPADE."	ט	ط t	⌒ d
"20," "CAH." "A SLING."	כ	ك , ك q	⌣
"40," "MIM." "WATER."	מ	م m	🦉
"60," "SAH." "THE MOON."	ס	س s	─╫─
"80," "PHEH." "THE FACE."	פ	ف f	□
"100," "KAV." "A BOAT."	ק	ن k	Δ q

Figure 7:
Hebrew, Arabic and Hieroglyphic Symbology of the Alphabet, 2/4

Number (Lamb, p.12 +)	Hebrew	Arabic (Whallis, p.cli)	Hieroglyphic (Griffith, p.15)
"300," "SHISH." "THE SUN."	שש	ش sh	 ⏣ š (sh)

Hieroglyphics
that are missing
\\ i cutting
⌐ b foot
∽ f sand viper 
◯ r libia majora circomcised lips

Figure 7:
Hebrew, Arabic and Hieroglyphic Symbology of the Alphabet, 3/4

Number (Lamb, p.12 +)	Hebrew	Arabic (Whallis, p.cli)	Hieroglyphic (Griffith, p.15)
"1," "AD." "A MAN."	א	ا `	 𓅭 (a)
"2," "BETH." "AN HOUSE."	ב	ب b	
"4," "DAU." "THE LIPS."	ד	د d	
"6," "VAV." "A FEATHER."	ו	و w	 𓅱 w (u)
"8," "CHACH." "THE BOSOM."	ח	ح ḥ	𓎛
"10," "AI." "THE EYE."	י	ي y	𓏭 y (á)
"30," "LI." "A LION."	ל	ل l	
"50," "NAH." "A CUP."	נ	ن n	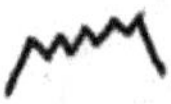 ∿∿
"70," "AUL." "THE LEGS."	ע	ع '	 ⌒ (á)

244

Figure 7:
Hebrew, Arabic and Hieroglyphic Symbology of the Alphabet, 4/4

Number (Lamb, p.12 +)	Hebrew	Arabic (Whallis, p.cli)	Hieroglyphic (Griffith, p.15)
"90," "TZI." "A HORNED BEAST."	צ	ص ṣ	
"200," "RAH." "A HAWK."	ר	ر r	
"400," "THATH." "A TENT."	ת	ت t	t
"500"		ث th	ṯ (ṯh)
"600"		خ kh	ḫ (ḵh)
"700"		ذ dh	ḥ (ch)
"800"		ض ḍ	
"900"		ظ ẓ	z
"1000"		غ gh	

Chapter 35: Conclusions

What gives me a right to make this book to be so? According to the Constitution of the U.S. of America, in Article 1:

> Congress shall make no law respecting an establishment of religion, or prohibiting the free exercise thereof; or abridging the freedom of speech, or of the press; or the right of the people peaceably to assemble, and to petition the government for a redress of grievances. (Rawle, p.340)

I believe, in my understanding and estimation, in this Article 1 of the U.S. Constitution, the concept of peace in protest, giving your free speech, letting people find their own religion and spiritual path and using the free press to publish ideas. That is what this book is from, free ideas to the press.

If anyone wants to vote for me at the 2028 U.S. of America Presidential Election on November 7, 2028, then write-in Matt Mandell. I can find a vice president after I'm elected, because I can put this into the **+whirlwind+** for people to subconsciously know.

I hope to **+Trot+** into a peaceful spiritual revolution with this book so that we can counter evil forces and **+Say-tin.+** Can you say Leon Trotsky? Satin is the evil, similar to the Soviet Union, a force that can lead people astray as in the **+snakehead+** +S-M.+ I would think that in this age of information, that the information of the poor and oppressed will also come forward and be heard as a consequence of this information revolution we are entering today and into the middle twenty-first century.

Listening to SV from the searching of intuition in CV will lead to the opposite in SV. Because the CV intuition is transformed by a reflection back into SV to give the opposite. I think this is the reason why I'm losing so much money trading ES-futures. I would trade ES-Futures using SV to do what my **+Meself+** says, I would focus on the short term and miss the long term trend because I would be trading against the long term market movement. ~**It seemed logical but there are ogres.**~ ~**Matt when you try to talk and see which way ES will go get the opposite. Hurry back and jettison and you can win Matthew. Matt when you put your intuition in you get the opposite.**~ So I should trade when I get the insight and place the trade, then do SV afterwards!

My concept of our Universe is based on the idea that our Universe is bileveled by way of its dimensions. The dimensions can transform length into inverse length. This is because the universe has opposites. Regular Mass goes into the black hole and dark mass comes out. There is a balanced equation to the universe that requires balanced energies and masses.

People do not know who leads their thoughts because they cannot consciously see their thoughts. By doing SV people will be astounded, and bootstrap their natural God-given intuition, to the fact that Satin exists today and is probably telling many people what to do, without them knowing it. I will lead this revolution by uncovering, talking about this book I typed on my computer and take it as an active role in this debate where I can find it or make it.

The world is in an unconscious matrix, for example we don't really hear the actors on a movie screen speak, because the audio is in an audio track for the film. Hence, when we act with our actions, for example, build a bench with wood and the thoughts that we use to make the bench itself are not observed. Since we cannot see thought, similar to the audio being separated in a movie film, then is it possible that thought is just of spirit, partly in the 5D too. To make reality, the electrical activity in the brain, in the R3DT, is a result of the spirit being there as well as the soul. The soul is recording the deeds fulfilled in one's life.

When we say a word with CV, it is in the image of God. I think this is so, because those thoughts are partly from the spirit so they make a word come out of the mouth. The thought that instigates a word, as processed through the brain from the spirit, is superseded from the unconscious components of symbols of truth, as light shining through a completed image from a stained glass window. Be careful what you look at or let into your eyes.

The gray squirrel, when you look at their eyes can go into your **+whirlwind,+** from my experience. **~The squirrel can go into your whirlwind, through your eye.~** This is August 10, 2024 my last paragraph, the end.

When someone is pumped-up in thoughts, kind of similar to when a price is pumped-up for a stock and then dumped, then the spirit pumping up one's thoughts are leaving the crime to the individual holding the evidence and this crime is at this individuals soul, who physically did it in the R3DT. Hieroglyphics of shame can be written on this individual's soul that physically did it or didn't do the right thing.

In ancient Egypt, it was a society that was not fearful to be sexual, but this lead to shame. I think, ancient Egypt was a society that made men and women equal, this was harmony, but they sexualized the reproductive body parts, equal men and women, as I have seen when doing my interpretation of some of their hieroglyphs based on my cultural biases. My **+Meself+** said, **~The ancient Egyptian lives were horrible, the men took advantage of the women.~** I guess what I said was wrong. I ask, "Did Pharaoh make the lives horrible with slavery?"

I think that the **+parallax+** is a real thing outside the human body. Since God created humans in his image, I believe that yes, this is because the universe would also be in the image of God, because he used his word to make it be. God is the word, the first and the last morphed together and this is similar to the **+parallax.+** This is as a ring, connected as a big loop's ends put together as one and twisted to make the infinity symbol, as a **+whirlwind+** in its circuit.

World Peace can happen if we have the same language, telepathy. We do telepathy by putting the two into one, read Lambdin, et al. The frequency, 1/s, into the time, s, and then multiplied together gives s/s or 1. The two being the conscious "I" and ego and the other the **+meself+** or spirit. The conscious is electric and the subconscious is magnetic. If we can put these two thoughts together and make light, as composed of electric and magnetic waves together, then the ego will fall out. With the "all" we can move that mountain as Jesus suggests, from Lambdin, et al.

That mountain is just **+Mounta-Zumba.+** This is just "Zumba Rock" in Nigeria, according to a search on duckduckgo.com and not a big mountain. So if **+Mounta-Zumba+** is just a big rock that was a big mountain then let us all make the big rock into a zero dimensional point. Then when we say move, that point will move and it will move when we hear it in SV.

One can start by listening to SV, become mindful of their thoughts and filter them to be combined with the subconscious thought waves, ga, (Egyptian Hieroglyphic folding your hands to symbolize the all or one, number 3, from page 242) the two into one, from reading Lambdin, et al.

Baa-ah-ah I'm conducting a victory on the subconscious. Oh-oo-oh-ah taking the insides and pulling it to the outsides, finding the above and bringing it to the below.

Figure 8:

Testing you and the **+whirlwind+**

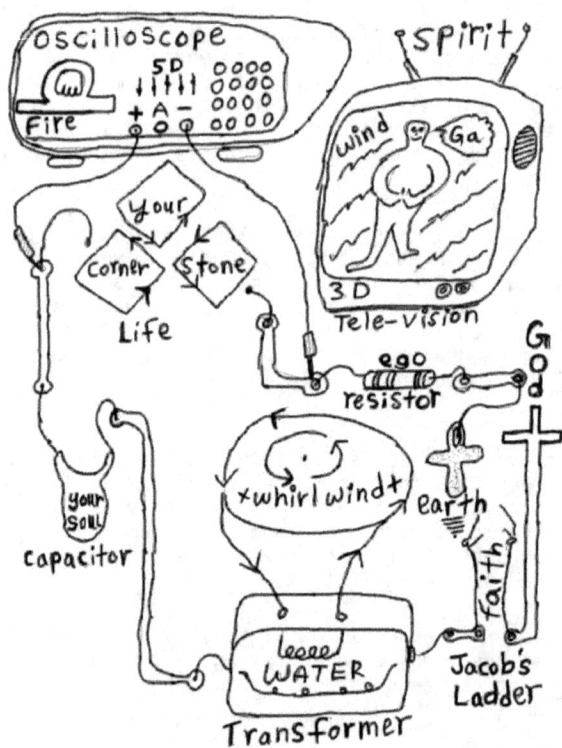

Appendix 1:

Below is my signed letter to organizations when I was handing out my business cards from http://the2SRVin.us website, (*).

Matt Mandell
40 Arlington Ave E
17433
St. Paul, MN 55117

Date: September 19, 2022

Your Organization or Home
Minneapolis and St. Paul
with surrounding communities

To Whom It May Concern,

In the past few weeks I have visited many business, churches and people. I was trying to deliver my business card because I thought I could help as a counselor to some people. I realize that some may think I will be replacing their teachings and programs. I think that this is just opposite to the truth. I believe my subconscious analysis is not to replace, but to only be a tool to support it as I would love to talk to you to see if this is so.

I'm writing to you to introduce, what I think is, a way to access the subconscious and is only a matter of time before this will be accepted as a valid way to do so. If one is able to [do] their backwards audio while also believing in its theory and then that this has a usefulness then they can contemplate information from their higher self that can change their bad behaviors or habits to become more successful in life. If parents to a child, for example, know that their child is not learning well then my subconscious analysis can attempt to open, to the child with the parents consent, the higher self.

The higher self can offer insights and tips as to what is going wrong or right. If this child can also listen and hear these metaphor and story-lines themselves then it will be conformation from the subconscious that they are connected to a higher self and this information can be a help with their conscious understanding of what they are doing and to be more self aware. Increasing self awareness is the key to increasing the likelihood one is able to surmount a problem or complete a goal. Problems can be solved and goals won by understanding the end state and this is what the subconscious can reveal.

I look forward to finding clients that I can be a
volunteer listener to, for free, by recording their
spoken words describing what topics or interests
they are concerned with. Then taking this recording
and listen to it backwards while then identifying
the insights from the subconscious. The audio could
be sent to me in .mp3 format in email or with a web
camera conference and recording it that way. I could
also listen to clients and ask question in person
and record with my hand held voice recorder and then
send the results to the client's email, if the
client is within the Twin Cities.

From the thoughts of,
Matt Mandell
http://www.The2SRVin.us

Appendix 2:

Below are my SBTM/BTM/SV postulates originally on my
website from source (*).

BTM Postulate 1:

3D and 5D are separated by the firmament with
the spirit of their waters also separated,
this gives a distinction between space and
thought. The +whirlwind+ is the energy of a
place or thought put into not the spirit but
the water and this is how we have telepathic
ability or to transfer between the 3D and 5D.
Therefore, frequencies of space and thought
can also separate and combine. Like
frequencies combine and unlike frequencies
separate or repel into a non-event.
version: 7162022+1

BTM Postulate 2:

The third eye or pineal gland, when it is
unclogged, can see the +whirlwind+ in other
people or places and things. The result is
that one's body will feel this as intuition.
3D is coupled to time, and therefore in
situations with unlike frequencies the result

is one will want to repel and the someone
will want to move in a space away. 5D is
timeless and therefore unlike frequencies
will cause a nightmare and one will get away
by waking-up, to get away from a timeless
encounter.
version: 7162022

BTM Postulate 3:

Since 5D is timeless then the future is
created in the 5D and this future will
transport backwards in time to the present,
into the 3D, to make reality in our
[U]universe. Spaciously, an event can happen
when the future frequency of space matches
the present space frequency. If these
spacious frequencies don't match then they
will subtract from one another and move into
an alternate universe where they do add
together. As a result one cannot physically
go back into time to change the future of
themselves.
version: 7162022

BTM Postulate 4:

Since events happen backwards in the
universe, starting in the 5D, this means that
causality in the universe will occur because
the frequency of spaciousness and
thoughtfulness occur[,] before they are
realized consciously in the present 3D.
Messages from BTM will go backwards from the
forward speech in 3D as a result of the
subconscious Self being of spirit is residing
in the 5D.
version: 7162022

BTM Postulate 5:

Free will allows for the thought to be in
time or timeless and able to be uncoupled
from space. As a result one can imagine what
happened in the past and go there remotely in
a thought. Thought is in existence outside

the universe however this is not always true
for space. Space in our 3D is not the same
outside the universe because the 3D was
constructed by God from raw materials. The
configuration of the 3D is not necessarily
constructed the same as ours outside the
universe.
Version: 7182022

BTM/SV Postulates Conclusion:

What is the conclusion of the 5 Postulates?
The Soul is not the +Me+. At death the +Me+
is a zero because it was the Ego in that life
and the Soul now considers it a past life.
The Soul is not both space and thought. After
one passes away, the Soul is released from
the body. The Soul energy was from the higher
dimensions and at death that is where it
returns. So then after death maybe the Soul
can go to the 5D with the +Meself+. I believe
that the +Meself+ then resides in the 5D and
is comparable to the Holy Ghost, only because
we are in the image of God. While in the 5D
the Soul can decide to start again and be the
"Father" to the "Son" in the new life.
Probably the past life can help the next life
learn through the concept of karma from the
past lives.
Version: 9292022

Appendix 3:

Below are my ten objectives to get done if elected
to the Office of the President of the United States.
These are my ideas as, Matt Mandell for President,
originally on my website from source (*).

Platform
1) I have no committee for the FEC
-write-in, Matt Mandell
-looking for a VP
2) Take National Issues not States'
-States should fend for their issues
-Congress should help States
3) Reduce Federal Taxes

- Color coded credit cards for ones' tax rate, from 0 to 20%, pay-as-you-go
- Keep current system grandfathered

4) National Volunteer Police, NVP
-Connect the dots from federal to local
-Require gun and rifle ownership to have three sponsors

5) Coordinate with National Agencies for: Harvest, Intelligence, Validation, Express, HIVE
-Connect the dots
-Disseminate to the People

6) Be Energy Independent
-Find ways to promote and use energy
-Export to other countries

7) Be Job Independent
-Promote jobs and innovations
-National database for the unemployed and the skills and experience of job seekers

8) Reduce Government
-Find ways to streamline costs and dept
-Care for the environment and export its resources Nationally and World wide

9) Coordinate with other countries for World Peace
-Connect and show security and safety
-Show commonalities and draw boundaries

10) First Contact
-In, on and in orbit of the Earth
-Alien contact could be to our benefit
Confirmed on July 25, 2022

Appendix 4:

Installing the Ubuntu Studio operating system so you can have an audio editor on your computer.

You should have a flash USB drive that is at least 16GB in size, because you will use this to burn a bootable image of Ubuntu Studio onto it. The website is ubuntustudio.org and I will list the steps I took to do this below.

1) Assuming you have an Intel or AMD CPUs then download its latest version, Ubuntu Studio 22.04.2 LTS to your downloads folder. This

should download one operating system in a *.iso file (the * is the exact filename) into your download folder with your internet browser, if not I know Chrome works well to do this. Otherwise, you will get an ark-folder and you might be able to live install it from there. This might be helpful if you have a Microsoft Windows computer.
2) Next, launch Terminal or Konsole to navigate to the Downloads folder and check that the SHA256SUMS file matches the key when you input the command, sudo sha256sum *.iso to get the hash check which should be the same as the download website's hash check.
3) Apparently there is no way to be sure if you have a good *.iso unless you check the identification of the author in the keyring, in person through the procedure, outlined on the Ubuntu website.
4) Next make a *.iso image onto the USB drive with the commands:
>   *lsblk* to find the name of your USB drive, might be sdc, then:
>   *sudo dd if=*.iso of=/dev/sdc status="progress" bs=1M*
5) Next turn off the computer and boot to the USB drive by plugging-in the USB drive into a USB port, turn on the computer, and press the function key to get to the boot menu or just select to install the Linux system you wanted from the automated prompt.
6) I recommend choosing encryption, then your user name, password and select install.
7) After you log-in for the first time you should load Gufw, a firewall program, turn it on, and then load the software updates.

Appendix 5:

Below is part of my signed letter for a Freedom of Information Act Request (FOIA) sent to the Defense Intelligence Agency's FOIA mailing address on December 16, 2023.

FOIA: Freedom of Information Act Request

FOIA Requester Service Center
Defense Intelligence Agency (DIA)

ATTN: IMO-2C (FOIA)
7400 Pentagon
Washington, DC 20301-7400
Date: December 16, 2023

Dear: To Whom It May Concern,

This is a request under the Freedom of Information Act (5 U.S.C. § 552).

I request that a copy of the following from the DIA be sent to me: document, survey, e-mail, memorandum, in printed or electronic format e.g., PDF or WORD, containing any one of the phrases, "reverse speech" or "backwards speech" also "David John Oates" or "D. J. Oates" this could be with the subject or title about a(n), intelligence analysis brief, employment screening, subconscious study, interrogation technique, forecasting of the future, channeling of a past event or psychic experiment.

This request is not intending to include documents etc. associated with reverse phonation or having reverse speech used as a control group or condition in a scientific experiment. Because this is not primarily interested in reverse speech as a way or method to access the subconscious or unconscious. In order to help you determine my status for the purpose of assessing fees, you should know that I am an individual seeking information for use in my manuscript I intend to publish, *2into1.com SV: a +reverse speech+ voice for World Peace*, and my website I intend to put on the internet, all for commercial use with the Creative Commons License: CC BY, information at creativecommons.org.

I am willing to pay fees for this request up to a maximum of $25. If you estimate that the fees will exceed this limit, of $25, then please inform me first.
[. . .and signed, dated and with my return address.]

There has been no response to my FOIA letter from the U.S. Government about reverse speech before September 17, 2024. If there is a response then I will post it on 2intone.us . . . .

Suggested Reading:

American Hypnosis Association, (AHA). (n.d.). Reverse Speech Technology Proof of a Duality of Consciousness. [Hypnosis Motivation Institute]. Retrieved from https://hypnosis.edu/aha/articles/reverse-speech

Deutsch, D. (2019). *Musical Illusion and Phantom Words*. New York, NY: Oxford University Press.

Dunning, B. Skeptoid.com. (June 17, 2008). When People Talk Backwards. [Podcast]. Retrieved from https://skeptoid.com/episodes/4105

Example.com. (n.d.). 7+ Reverse Speech Examples - PDF. [Online knowledge base]. Retrieved from https://www.examples.com/business/reverse-speech.html

Kaufman, F. Vidmore.com. (July 30, 2020). Top 5 Backwards Recorders to Reverse Audio on iPhone/Android/Online. [Software company]. Retrieved from https://www.vidmore.com/record-audio/backwards-recorder/

Newbrook, M., & Curtain, J. (1998). Oates' theory of reverse speech: an update. *The Skeptic*, 15 - 16.

Newbrook, M. (2005, June). Reverse Speech., 15.2. Retrieved from https://skepticalinquirer.org/newsletter/reverse-speech/

Oates, D. J. (1991). Reverse Speech, Hidden Messages in Human Communication. Indianapolis, IN: Knowledge Systems, Inc.

Oates, D. J. (2022). The Great Reverse Speech Lexicon. Hackham SA, Australia: Reverse Speech Pty Ltd.

Skepdic.com. (November 24, 2015). reverse speech. [Online knowledge base]. Retrieved from https://www.skepdic.com/reversespeech.html

Stollznow, K. (2014). Reverse Speech. In: Language Myths, Mysteries and Magic. Palgrave Macmillan, London. https://doi.org/10.1057/9781137404862_14

References:

American Bible Society. (1858). *The Holy Bible : containing the Old and New Testaments* [DX Reader versions]. Retrieved from https://archive.org/details/holybiblecontain00amer_4/page/n8/mode/1up

Bonini, L., Rotunno, C., Arcuri, E., & Gallese V. (2022). Mirror neurons 30 years later: implications and applications. *Trends in Cognitive Sciences.* 26(9), 767-781. Retrieved from https://doi.org/10.1016/j.tics.2022.06.003

Byrne, T., & Normand, M. (2000). The demon-haunted sentence: A skeptical analysis of reverse speech. *The Skeptical Inquirer,* 24(2), 46–49. Retrieved from https://scholarlycommons.pacific.edu/cop-facarticles/264

Catholic Software, (1999). *The Bible, Douay-Rheims, Old Testament—Part 2.* Urbana, Illinois: Project Gutenberg. Retrieved from https://www.gutenberg.org/ebooks/1610

Griffith, F. (1898). *A collection of hieroglyphs; A contribution to the history of Egyptian writing* [DX Reader version]. Retrieved from https://archive.org/details/collectionofhier06grif/mode/1up

Hosseini, E. (2021). Brain-to-brain communication: the possible role of brain electromagnetic fields (As a Potential Hypothesis). *Heliyon.* 7:3, 1-9. Retrieved from https://doi.org/10.1016/j.heliyon.2021.e06363

Lamb, J. (1835). *Hebrew Characters; Derived From; Hieroglyphics. The Original Pictures Applied To The Interpretation Of Various Words And Passages In The*

Sacred Writings; And Especially Of; The History Of The Creation And Fall Of Man. Second Edition. To Which Is Added; An Inquiry Into The Origin And Purport; Of The; Rites Of Bacchus [DX Reader version]. Retrieved from https://archive.org/details/hebrewcharacters00lambuoft/page/n11/mode/2up

Lambdin, T., Grenfell, B., Hunt A., Layton, B., & Schenk, C. (1992). *The Gospel of Thomas.* Retrieved from https://sacred-texts.com/chr/thomas.htm

Mandell, M. www.The2SRVin.us, www.The2SRVin.us LLC., 2021, http://The2SRVin.us, Last access approximately May, 2022.

Ooguri, H., & Yin, Z. (1997). *TASI Lectures on Perturbative String Theories. arXIV: High Energy Physics - Theory (1996): n. Pag.* Retrieved from https://arxiv.org/abs/hep-th/9612254

Pauen, M. & Haynes, J-D. (2021). Measuring the mental. *Conscious and Cognition.* 90:1053-8100 (2021) 103106 Retrieved from https://doi.org/10.1016/j.concog.2021.103106

Rawle, W. (1829). *A view of the constitutions of the United States of America* [DX Reader version]. Retrieved from https://archive.org/details/viewofconstituti00rawl/mode/1up

Wallis, E. (1920). *An Egyptian Hieroglyphic Dictionary. With An Index Of English Words, King List And Geological List With Indexes, List Of Hieroglyphic Characters, Coptic And Semitic Alphabets, Etc.* [DX Reader version]. Retrieved from https://archive.org/details/egyptianhierogly01budguoft/page/n6/mode/1up

Index:

alien(s) 3, 9, 47, 56, 111-2, 121, 128, 148, 175-6, 218, 234, 255
apophenia 14, 114

Artifical Intelligence, (A.I.) 11, 14, 140, 133, 157, 174-6, 185, 202-4
astral 36, 68, 88, 94-5, 103, 177, 179, 205, 213, 220, 236
audio 8, 12-5, 17, 31, 53, 60, 69-71, 81, 94, 98, 100, 105-6, 123, 130, 138, 140, 146-7, 202-3, 205, 222, 226, 230-1, 247, 251-2, 255, 258
backward speech 14
bilevel(s) 10, 32-3, 42-3, 51-2, 58, 165, 190, 247
brain(s) 3, 8, 14, 17, 20, 30, 44-5, 53-5, 57-63, 65-6, 69-72, 80-3, 85, 89, 99-101, 104, 107, 109, 111, 113, 130-1, 135-8, 141-6, 150-8, 165, 169, 175, 184, 186, 195, 198-9, 202-3, 205-6, 208-9, 225-6, 229, 231-2, 234-5, 247-8, 259
cerebellum 8, 44, 54-5, 70, 89, 133, 143, 153-4, 156-7, 166
cerebral cortex 44-5, 89, 133, 151, 154, 157, 204
chakra(s) 15, 36, 53, 60, 72, 85-7, 95, 101, 114, 122, 124, 126, 134, 136, 140, 145, 158, 161, 163, 183, 194-5, 197, 200-1, 204-6
Cryptochrome-2 198-9
David John Oates 13, 257
diabetes 9
diet(-) 7, 61, 74-7, 101, 206, 212
DNA 11, 102, 148, 157, 181-2
dream(-) 1-2, 4, 7-8, 15, 21, 27, 30-1, 35-6, 38, 41, 53, 58, 60, 65, 70, 74-7, 79-82, 86, 95, 97, 100, 105, 107, 110-2, 127-8, 135, 138, 143, 145-7, 153, 161, 165, 170, 177-80, 207, 209-16, 220-2, 124, 226, 229, 232, 237-8
Egypt(-) 1-2, 4, 9, 19, 128, 213, 237-41, 248-9, 259-60
Fibonacci 91, 108, 170, 184-5
firmament 15, 36, 177, 189, 218, 221, 229, 252
-fluorid- 148, 157, 178, 207
garden 38, 127-9, 187, 190, 205, 239
God(-) 11, 13, 18, 36, 103, 117-8, 120, 123, 128, 149, 157, 180, 187-8, 190, 199, 202, 208, 218, 220, 225, 227-8, 235, 237, 241, 247-8, 254
"Gospel of Thomas" 1, 29, 224, 227-8, 260
Grand Unified Theory (GUT) 1-3, 15, 20, 27, 34, 42, 45, 60-1, 89-90, 93-4, 183, 191, 222, 259
hallucinations 231
hidden dimensions 20-2, 32, 42-4, 49-50, 88
hypnosis 58, 60, 100, 130, 137, 258
image(s) 8, 13, 15-6, 18, 45, 58, 63, 100, 120, 134, 140, 157, 164, 173, 202, 207, 212, 219, 229, 231, 247-8, 254-6

imaginary 7, 23-5, 31, 39, 42, 47, 50, 181, 191-3, 215
intuition 7, 14-5, 29, 55, 60-1, 64-5, 72, 79, 86, 96, 109, 114, 124, 126, 134, 136, 142, 156, 167, 172, 179, 187-8, 198, 204-5, 208, 217, 219-20, 222, 226-7, 229-31, 235, 246-7, 252
linear 19-20, 26, 28, 32, 35, 42-5, 58, 127, 129
meditation 18
memor(-) 8, 31, 53-60, 63, 68, 75, 80-1, 89, 99, 104, 114, 116, 130, 135, 140-, 144, 146, 154, 219, 226, 230-1, 257
metacognition 2, 104-5, 182
mind- 3, 8-10, 12, 14-5, 18, 29-30, 38, 53-4, 60-6, 68-9, 80, 85, 97, 99-100, 104-5, 107-9, 111, 113-6, 120, 122-3, 125-6, 128-34, 138, 143, 146, 149-50, 156-9, 164-5, 167, 169, 173, 178, 181-2, 187, 197, 207-8, 229, 233, 249
+Mir-re+ 8, 13, 163, 234
Mirror Neurons (MN)...........................198-9
+ogre(s)+ 19, 38, 65-6, 85-7, 97, 109, 140, 149-50, 155-8, 164, 169, 184, 191, 198, 209, 221, 247, 256
palindromes 18, 92, 113
+parallax+ 3, 19-2, 44, 72, 82, 152, 248
past li(-) 4, 116, 132, 199-200, 202, 221, 254
personal mythology 181
physical(-) 33, 50, 55, 69, 90, 104, 119, 145, 149, 191, 202, 218, 248, 253
pineal gland 4, 7, 44-5, 83, 86, 89, 99-103, 133-4, 137, 151, 155, 178-9, 181, 201-2, 204-7, 209-10, 222, 226, 235, 252
programm- 3, 23, 45, 61, 76-7, 115-9, 131
Repressed Memory Regression 55
reverse back-masking 2, 12, 14, 98
rotate(s) 10, 22, 27, 95, 188-9
sense 20, 26, 30, 39, 50, 53, 58, 60, 64, 68, 73, 75-6, 80-1, 83-5, 91, 114, 116-8, 135, 141, 143, 145-6, 158-9, 169, 172, 191, 201, 203-6, 211, 220-3, 235, 240
soul(s) 2, 7, 9, 13, 16, 64, 66, 68, 80, 96-7, 102, 117, 160, 184, 189-91, 199-202, 205-6, 208, 214, 220, 223-5, 227, 229-30, 247-8, 254
source(s) 2, 7, 35, 69, 71-2, 80, 85, 91-4, 98, 102, 118, 132, 144, 146-8, 154, 158, 168, 176, 189-90, 205, 208, 232-3, 254-5
spirit- 2, 8-9, 11-3, 15-8, 36, 40-1, 70, 88, 92, 96, 101, 122, 135, 161-3, 166, 168, 182-3, 190-1, 199-202, 208, 216, 218-21, 223-7, 229-32, 235-6, 239, 246-9, 252-3

starfriend- 9, 15, 124, 234
story-line- 3, 12, 15, 17, 19, 31, 59, 71, 73, 81-2,
92, 94-6, 98, 102, 106-7, 109, 114, 120, 130, 133,
138, 140, 147-9, 154-5, 160-1, 165, 169-73, 175-6,
181, 187, 202, 208, 230-1, 251
String Theor- 15, 27, 30-1, 39-42, 91, 129, 178,
191, 194, 236, 260
subconscious- channel- (SC) ...2, 14, 71, 191, 206-7
sugar 9, 10, 101, 212
symbol- 1, 3, 8, 13-5, 38, 58-9, 61-3, 75, 80-2,
104, 107, 109-10, 120, 125-6, 128, 135, 142-3, 171,
196, 207, 242-5, 248-9
timeless- 15, 20, 22, 30, 32, 35-7, 39, 42-4, 48-9,
52, 74, 95, 127, 155, 159-61, 172, 200, 219, 229,
231, 253
Thoth 237-8
two-step 53-4
Ubuntu Studio 255
World Government..............................9, 168
worm hole 22, 35-6, 42, 91-2, 127-9, 214
zero point- 8-12, 15, 17, 20-23, 25, 29, 32-4, 42-5,
48-9, 52, 127, 170, 172, 189, 215

www.ingramcontent.com/pod-product-compliance
Lightning Source LLC
Chambersburg PA
CBHW052245220526
45471CB00001B/195